D1065473

DISTILLING DEMOCRACY

DISTILLING DEMOCRACY

ALCOHOL EDUCATION IN AMERICA'S
PUBLIC SCHOOLS, 1880-1925

Jonathan Zimmerman

University Press of Kansas

© 1999 by the University Press of Kansas
All rights reserved

Published by the University Press of Kansas (Lawrence, Kansas 66049), which was organized by the Kansas Board of Regents and is operated and funded by Emporia State University, Fort Hays State University, Kansas State University, Pittsburg State University, the University of Kansas, and Wichita State University.

Library of Congress Cataloging-in-Publication Data

Zimmerman, Jonathan, 1961–
    Distilling democracy : alcohol education in America's public schools, 1880-1925 / Jonathan Zimmerman.
        p.   cm.
    Includes bibliographical references and index.
    ISBN 0-7006-0945-8 (cloth : alk. paper)
    1. Alcoholism—Study and teaching—United States—History. 2. Temperance—United States—History. I. Title.
    HV5128.U5Z55 1999
    362.292′071′073—dc21   98-44935

British Library Cataloguing in Publication Data is available.

Printed in the United States of America
10 9 8 7 6 5 4 3 2 1

For Susan, the Real Doctor, with love

CONTENTS

ABBREVIATIONS OF MANUSCRIPT COLLECTIONS

Alfred WCTU	Woman's Christian Temperance Union of Alfred (N.Y.) Records, Alfred University Library, Alfred, New York
Atwater Papers	Wilbur O. Atwater Papers, Olin Library, Wesleyan University, Middletown, Connecticut
AU Papers	American University Presidential Papers, American University Library, Washington, D.C.
Council of School Superintendents	Records of the Council of School Superintendents, Cities and Villages, State of New York, Department of Manuscripts and University Archives, Cornell University Library, Ithaca, New York
Dunham Papers	Catharine Dunham Papers, Arthur and Elizabeth Schlesinger Library, Radcliffe College, Cambridge, Massachusetts
Michigan WCTU	Woman's Christian Temperance Union of Michigan Records, Bentley Historical Library, Ann Arbor, Michigan
Skinner Papers	Charles Rufus Skinner Papers, Jefferson County Historical Society, Watertown, New York
STF Series	Scientific Temperance Federation Series, Temperance

	and Prohibition Papers, Ohio Historical Society (joint Ohio Historical Society—Michigan Historical Collections)
STF Papers	Scientific Temperance Federation Papers, New York Public Library, New York, New York
Vermont WCTU	Woman's Christian Temperance Union of Vermont Records, Bailey-Howe Library, University of Vermont, Burlington, Vermont
Welch Papers	William H. Welch Papers, The Alan Mason Chesney Archives of the Johns Hopkins Medical Institutions, Baltimore, Maryland
Wilder Papers	Burt Green Wilder Papers, Department of Manuscripts and Archives, Cornell University Library, Ithaca, New York

PREFACE

I conceived of this book over a decade ago, while teaching social studies at a public high school in Vermont. I also taught "drug and alcohol education," a state-mandated unit on the dangers of these substances. Midway through the year, I began to detect a sharp tension between these two duties. My central aim in teaching was—and is—to develop my students' critical, deliberative capacities. Yet the drug-and-alcohol curriculum was not open to deliberation. Instead, the state required me to transmit a single, unequivocal truth: that all illegal drug use and underage drinking was harmful. To question this proposition in front of students was to engage in "enabling," I was told. It was also to risk losing your job.

I'd like to say that I broke the law and took the heat, sacrificing my first teaching position to the larger cause of truth, democracy, and freedom. Instead, like most of us, I played along. Did my students sense my skepticism? Probably. Did they share it? Definitely. Every kid in the class knew somebody who was "messed up" on drugs or alcohol. But they also knew countless others—peers as well as parents—who used these substances in moderation, without any lasting ill effects. So they could not believe lessons on the so-called gateway effect of marijuana, for example, whereby pot leads inexorably to the harder stuff. When I described this specious "effect," they smiled—and I cringed. For theirs was the smile of the cynic, of children who are learning to dissemble like adults.[1]

Several years later, I found myself in graduate school, casting about for a dissertation topic. Suddenly I recalled my encounter with drug and alcohol education, which seemed ripe for that special brand of muckraking that historians prize. By probing the origins of this subject in public schools, I would show how America had gone so very Wrong. No less importantly, of course, I would also show how I was—or came to be—so very Right.

Having preestablished my conclusion, in other words, I set out to prove it. At first glance, it promised to be simple. I already knew that American education underwent a massive overhaul around the turn of the century, when centralized city school boards began to replace local, ward-based trustees. State systems burgeoned as well, bringing remote rural districts more firmly into the bureaucratic orbit. At the top of these systems stood a small cadre of experts, an "Educational Trust" of superintendents, professors, and university presidents who dominated school policy for the next few decades. To start my research, then, I looked first to this elite core of leaders—men like Charles W. Eliot, Nicholas Murray Butler, and David Starr Jordan. If drug and alcohol education extended that far into the past, I reasoned, they must have played a big part in establishing it.

On both counts, I was wildly off base. Alcohol education dated all the way back to the birth of America's common schools, in the 1830s and 1840s, when students marched in "Gospel Temperance" parades and received sporadic lessons on the dangers of liquor. By the 1880s, I learned, states began to require a graded, systematic study of "the nature and effects of alcoholic drinks and other narcotics." Dubbed Scientific Temperance Instruction (STI), this curriculum was firmly in place at least a decade before the Educational Trust came to the fore. It originated not with expert policymakers but with a little-known Massachusetts housewife named Mary Hanchett Hunt. Streaming first into legislatures and then into classrooms, Hunt and her army of assistants in the Woman's Christian Temperance Union (WCTU) made STI mandatory across the country by 1901. From smoke-filled statehouses to one-room schoolhouses, in principals' offices and teachers' institutes, on Indian reservations and in military academies, even at schools for the blind and deaf, thousands of untrained and unheralded women brought Scientific Temperance to millions of American children.

In many ways, their message mirrored the one that I taught in Vermont. But they made alcohol the demon, investing it with the same evil qualities that we now impute to illegal drugs. Hunt and her aides insisted that alcohol was "poison," harmful to the human body in any amount. Textbooks featured lurid depictions of drunkards' stomachs, a schoolroom standby since the antebellum era. But even moderate drinking was said to damage the liver, heart, and nervous system. Indeed, like drug educators today, Hunt deemed "moderate" use impossible. "[I]t is the nature of a little of any liquor containing alcohol to create an appetite for more," she wrote. "The strongest warning should be urged against taking that little and thus forming the appetite."[2] As it is today, moreover, the strongest censure was urged against schools—and teachers—who failed to comply. In the classroom, Hunt argued, any equivocation would surely promote (we would say "enable") precisely the vice that STI was trying to stop.

All of these parallels merely deepened my conviction that I was Right. I had erred, of course, in presuming an expert pedigree for alcohol education. Yet I still despised the rigid dogmatism of Scientific Temperance, so close in spirit to our own drug education. So I resolved to frame my study as one of popular delusion, not professional malfeasance. The Educational Trust certainly bore much of the blame for intelligence testing, racial stereotyping, and a host of other evils that plagued our schools. In fairness, though, I could not saddle these experts with Scientific Temperance. It sprang instead from the intolerant soil of middle America, a poison weed with grass roots.[3]

Indeed, I learned, the experts denounced STI as vehemently as I opposed drug education. Even their objections resembled my own. Scientific Temperance was tedious as well as duplicitous, they argued, boring students with an endless litany of half-truths about alcohol. The lineup of STI's enemies reads like a veritable Murderer's Row of leading American educators in the early twentieth century. It included such heavy hitters as Eliot, Butler, and Jordan as well as Charles and Mary Beard, G. Stanley Hall, Francis Parker, and Alice Freeman Palmer. These experts disagreed sharply with each other on many school matters, of course. On Scientific Temperance, however, they stood united—in opposition. That was good news, I thought, yet further evidence that I was Right. How could STI have earned such a diverse set of foes unless it was Wrong?

Then, about a month into my research, I found an extraordinary letter. I still remember reading it, because it made me rethink this whole project. "If my child had scarlet fever, it would be the height of folly for me to call in a physician and demand that he cure him by the use of cod liver oil," wrote F. C. Atwell, a small-town New England school superintendent and a bitter critic of STI. "[T]hose who have . . . studied neither pedagogy nor psychology should be content to leave the details and the method of achieving the desired result to those who have."[4] To Atwell, Mary Hunt and her WCTU assistants were not simply "wrong." They were also quite literally out of line, trespassers onto a territory that was best left to professionals.

During these years, officials like Atwell increasingly took their cues from national school leaders. So I turned back to the Educational Trust, reexamining their attacks on Scientific Temperance. Here, too, I found a profound disdain for lay influence upon school affairs. "A school programme made by a legislature is about as reasonable as a dictionary or encyclopedia edited by the same high authority," quipped Nicholas Murray Butler, in a typical screed against STI. "The Grand Army of the Republic or the Knights of Labor might just as well get laws passed to further their aims and objectives."[5] Citizens like Mary Hunt had no business interfering with educational matters, even via their own elected representatives. Better to keep the barbarians outside the gates, lest schools suffer an endless avalanche of lay demands.

This was an argument about democracy, I came to realize, not about alcohol. The first time around, I was so charmed by the experts' critique of STI that I overlooked their dismissal of direct citizen action. But now, rereading their remarks with an eye to politics, I faced a fresh dilemma. How could I square my opposition to STI with my commitment to democracy? Back in Vermont, it had all seemed so easy. Drug and alcohol education inhibited democracy, spawning young cynics rather than deliberative citizens. Surely, though, it was popular. At PTA meetings, I recalled, substance abuse often topped the list of parental worries. National polls confirmed the same trend: in American education, only violence rivaled drugs as a source of public concern. Likewise, Mary Hunt's legislative successes as well as her legion of assistants bore tribute to the deep popular basis of Scientific Temperance. To dismiss all of this as a mass delusion placed me in league with F. C. Atwell, as his letter made me recognize. And that was not a place where I wanted to be.

So I resolved to find a new one, somewhere between expert and popular authority. On the one hand, expert directives seemed to deny citizens a basic, fundamental right to deliberate and transmit their values. As I learned in Vermont, however, a curriculum of "the people" could just as easily muzzle such debate. Did Scientific Temperance inhibit discussion, as well? After studying thousands of school reports, journals, newspapers, and letters, I can answer unequivocally: no. From Connecticut to California, STI brought hundreds of thousands of students, parents, teachers, principals, and superintendents into a multigauged debate about their common schools—and their common lives.

In taking this new tack, I hope I have not romanticized Mary Hunt and STI. At least five caveats are in order, lest I stand accused of embracing STI as cavalierly as I once dismissed it. First of all, I still reject its hidebound, formalistic approach to educating children. I come to bury Scientific Temperance, not to praise it. But in tracing its rise and fall, I try to show why dissidents like me must be willing to defer to popular educational movements—even those we despise. My goal here is simply to test the reader's faith in democracy, like STI tested mine. If this book sparks even a tiny fraction of the debate that once surrounded Scientific Temperance, I will count it a success.

Second, I want to make it clear that Mary Hunt and her assistants did not *intend* to provoke deliberation in American schools. Single-minded zealots, they sought to silence opposition—and squash discussion—wherever they saw it. But they failed. Scientific Temperance reminds us that even movements with repressive impulses can foster debate, given the proper context. Among the humanities, I like to think, history is uniquely suited for exploring and explaining such contexts. I hope this book bears me out.

Third, I must acknowledge that the STI debate rarely crossed lines of class, ethnicity, or race. Denouncing bans on alcohol, immigrant communities often stressed "education" as the key to limiting its use. So they hesitated to criticize Scientific Temperance, which seemed to reflect the same didactic emphasis they desired. Instead, STI's foes—like its friends—hailed almost exclusively from America's white Protestant middle class. Yet these rivals embodied distinctly liberal and conservative wings of that class. I hope, then, that STI can serve as a model of educational debate across social groups— not just within them.

Fourth, I do not mean to imply that Scientific Temperance was the only lay movement to influence the turn-of-the-century American curriculum. Although historians have rarely investigated them, dozens of other citizen groups also converged upon the classroom during these years. Butler's quip about "the Grand Army of the Republic and the Knights of Labor" might well have been ironic, since both organizations made strong efforts to alter school curricula.[6] In American schools, however, no group was as persistent—or as successful—as Mary Hunt's legion of WCTU women.[7] In our quest to understand lay influence upon the curriculum, then, Scientific Temperance seems like the best place to start.

Finally, I should also emphasize that Hunt and STI were fully implicated in the politics of American expertise. When I wrote the first draft of this book, flush with my "discovery" of Scientific Temperance, I cast it as a struggle between expert and popular authority—on one side the Educational Trust, on the other Mary Hunt. Yet Hunt and the WCTU often invoked expert authorities, just as their professional enemies invoked popular ones. Both sides, then, struggled to reconcile democracy and expertise. So do I.

The best part of completing a book is thanking everyone who contributed to it. The book began as a doctoral dissertation under the direction of Ronald G. Walters, a gifted mentor and a true friend. Ron's unstinting dedication to his students has set a standard that all of us seek—mostly in vain, I'm afraid—to emulate. I am also grateful for the wise counsel of Ellen Lagemann and Neil Postman at New York University, where they made me feel welcome from the start.

So many friends and colleagues read portions of this project that I cannot possibly list them all. But thanks especially to Laura Ahearn, Brian Balogh, Lewis Bateman, JoAnne Brown, Robert Brugger, John Burnham, Louis Galambos, Joseph Giaquinta, Richard Hamm, Floyd Hammack, Michael Katz, Edward McClellan, Howard Markel, Philip Pauly, Niko Pfund, William Reese, Dorothy Ross, Doug Rossinow, Margaret Rung, Donald Warren, Joel Westheimer, and Steven Wrinn. My dear friend and occasional office-mate Brett Gary provided good cheer and a careful critique of the Epilogue. I am indebted most of all to Joseph Kett, who believed in this book when my own faith was wavering. His generous advice and encouragement—at just the right time—generated the spark that I needed to finish it.

Nancy Scott Jackson brought the book to the University Press of Kansas, where she has guided it with subtle grace and skill. Thanks also to Rebecca Knight Giusti, Susan Schott, and freelance copy editor Carol A. Kennedy, who did much of the dirty work along the way. At NYU, I was further blessed with three superb secretaries: Delores Armstrong, Pam Baldwin, and Julie Jakolat. They made the arduous seem easy, and they kept me sane.

My other benefactors included the U.S. Department of Education, whose Jacob J. Javits Fellowship funded much of my graduate education; the Frederick Jackson Turner Society of the Johns Hopkins University Department of History, which helped finance several research trips; and my parents-in-law, Roberta and Larry Coffin, who donated one of the computers I used (and mangled!) in writing this book. I am also grateful to the editorial staffs of two journals for publishing parts of the project during its nascent stages. Portions of Chapters Two and Four appear as "'The Queen of the Lobby': Mary Hunt, Scientific Temperance, and the Dilemma of Democratic Education in America, 1879-1906," *History of Education Quarterly* 32 (1992): 1-30; of Chapter Three, as "'When the Doctors Disagree': Scientific Temperance and Scientific Authority, 1891-1906," *Journal of the History of Medicine and Allied Sciences* 48 (1993): 171-97; and of Chapter Five, as "The Dilemma of Miss Jolly: Scientific Temperance and Teacher Professionalism, 1882-1904," *History of Education Quarterly* 34 (1994): 413-31.

At the Frances E. Willard Memorial Library of the Woman's Christian Temperance Union in Evanston, Illinois, Alfred Epstein was an affable and knowledgeable guide. During several trips to Evanston I stayed with Tuck and Robbie McCollister, the world's most wonderful hosts. Robbie did not live to see the completion of this book, but I like to think that her spirit—one of friendship, tolerance, and respect—suffuses its pages.

From as early as I can recall, Margot and Paul Zimmerman taught me that ideas matter. Their care and dedication as parents made me—and this book—what we are. My two beautiful daughters, Sarah and Rebecca, brought me more joy than one man can possibly deserve. But my greatest debt is to Susan Coffin—brilliant doctor, dynamic mother, and devoted wife. She, alone, knows how much I love and admire her. The rest of you will have to take our word for it.

1

INTRODUCTION

In 1887, Minnesota became the nineteenth state to require Scientific Temperance Instruction (STI) in its public schools. Its law enjoined "all teachers" to provide "systematic and regular instruction" in the subject. It also designated a special committee—the state school superintendent and three normal-school presidents—to recommend "some suitable text-book" for teaching Scientific Temperance.[1]

Like other Americans, Minnesotans owed their STI law to Mary Hanchett Hunt and the Woman's Christian Temperance Union (WCTU). In accord with Hunt's "plan of work," local WCTU members held rallies and plastered their legislature with petitions for STI. After the bill passed, they also pressed the new textbook committee to adopt the "Pathfinder" books that Hunt endorsed. These texts deemed alcohol a "poison," invariably harmful to the heart, liver, stomach, and nervous system. Unlike the legislature, however, the textbook panel "discourteously disregarded" every entreaty of the WCTU—"the makers and movers and mothers of this Scientific Temperance Instruction Law," as Minnesota's WCTU president complained. The committee chose three alternative texts, condemning Hunt's Pathfinders as both inaccurate and inappropriate for young children. More importantly, it also rejected the "unwarrantable intrusion" of laypeople in the entire matter. Textbook decisions must be made only by "competent gentlemen," normal-school professor C. K. G. Hyde declared, not "by mere weight of numbers."[2]

Here Hyde provoked a stinging rebuke from Mary Hunt. "What you call the mere 'weight of numbers' is the final tribunal for all great questions under a government of the people," Hunt intoned. State WCTU officials agreed, calling on members to flood local school boards and block the committee's recommended books. Simultaneously, though, Hunt moved to assemble her own committee of "competent" Americans—behind *her* favored

1

books. Five months later, she released their names on a national petition. Signatories included "members of boards of health, influential State chemists, physicians, educators of national fame, the presidents of our great colleges and universities, and men known to the world as great ethical teachers," Hunt boasted. "As a whole, the petition constitutes an expression of the best expert sentiment of our country."[3]

This book explores how Hunt and her followers strove to reconcile their dual commitments to "the weight of numbers" and "expert sentiment"—that is, to democracy and professionalism. The same tension lay at the heart of Progressive-era politics and social thought in general, as a host of scholars have documented.[4] But that story is almost always told from the perspective of experts, not laypeople. Scientific Temperance Instruction shows us that citizens did not simply accept or reject professional authority, as many accounts would imply.[5] Instead, like the experts, they tried to use new knowledge to advance their own agendas. Neither passive victims of a deceitful seduction nor stubborn holdouts against modern expertise, Hunt and her followers had sound, self-interested reasons for selectively embracing it.

Indeed, they fused this new faith to a conservative philosophy of sin, free will, and personal regeneration. Historians have correctly associated such concepts with the "old" nineteenth-century middle class, which insisted that Americans could engineer their own success in this world—and their salvation in the next one.[6] Mary Hunt recast this attitude in a scientific idiom, but its essential structure remained the same. Evil lay in alcohol and in the people who chose to imbibe it; if students understood its true dangers, they could willingly cast it out. Here Hunt ran afoul of a growing liberal consensus among educators, scientists, theologians, and even other leaders in the WCTU. Deriding Hunt's emphasis upon personal autonomy, liberals claimed that vice stemmed not from depraved individuals but from a degraded environment. Hence these same foes of Scientific Temperance often endorsed alcohol prohibition, which generally targeted "the saloon"—that is, the social locus of drink—rather than liquor itself.[7]

Scientific Temperance, then, highlights both the convergence of and the conflict between two segments of the white Protestant middle class at the outset of our century. Most of the labels that historians use to describe this division—local versus cosmopolitan, traditional versus modern, rural versus urban, community versus society, and especially "old" versus "new" middle class—blind us to the critical element that both sides shared: a commitment to expert authority. So this book employs the terms "liberal" and "conservative," which connote a larger cultural difference in the way each wing perceived human nature. From small towns to great cities, in churches as well as classrooms, liberals emphasized the social roots of behavior: good and evil lay in the public square, not in the personal psyche. Conservatives, by contrast, stressed individual morality and character over social climate and causality. To Mary Hunt and her assistants, indeed, the only sure route to a better society was the mass conversion of its component members.[8]

This belief—*not* their attitude toward expertise—most clearly distinguished conservatives like Hunt from their liberal foes. Admittedly, Hunt commanded a dense army of lay assistants to enforce and defend Scientific Temperance. Such day-to-day citizen oversight contrasted sharply with the liberal ideal of public administration, in which credentialed specialists executed the common will. Even here, though, a shared faith in expertise bound STI's advocates and antagonists. Streaming into local schools, Hunt's assistants proudly proclaimed the names and claims of her own professional allies. Only with the death of Mary Hunt would this popular network of expertise dry up, signaling the rapid demise of Scientific Temperance Instruction in America.

Serving the People?
Expertise and Democracy in the Progressive Era

In 1910, historian Frederick Jackson Turner laid out a theory of expertise in democracy that would dominate his own profession for many years to come. "By training in science, in law, politics, economics, and history the universities may supply from the ranks of democracy administrators, legislators, judges and experts . . . who shall disinterestedly and intelligently mediate between contending interests," Turner declared. "It is hardly too much to say that the best hope of intelligent and principled progress . . . lies in the increasing influence of American universities." Unlike many of his contemporaries, Turner frankly acknowledged a division between "capitalistic classes" and "the proletariate" in modern America. Yet university-minted experts would stand above the fray, serving the entire People (a term Turner capitalized) rather than any special interest within it.[9]

For the next half-century, most historians seem to have accepted this benign interpretation of expertise.[10] In the 1960s, however, it came under heavy fire. Standing Turner on his head, some scholars condemned experts as the handmaidens of big business: from conservation and meat inspection to railroad regulation and workers' compensation, critics claimed, supposedly "neutral" experts did the bidding of corporate capital.[11] Other historians saw experts as an "interest" in their own right, using new forms of knowledge and organization to bolster their status. The key text in this group was Robert Wiebe's *Search for Order*, which tied the boom of experts to the birth of the modern state. At every level of government, Wiebe found, experts pressed for bureaucracies that demanded their technical skills and management. Doctors staffed city health departments, showcasing the germ theory of disease; state engineers inspected mines, reducing sickness as well as accidents; chemists tested patent medicines, which underwent federal scrutiny for the first time. Even the national forests were overseen by a fresh set of professionals, who selected trees for harvest as well as preservation. "It was the expert," Wiebe concluded, "who benefited most directly from the new framework."[12]

At the same time, Wiebe cautioned against explaining experts' rise to power solely in terms of their will to seize it. Borrowing from modernization theory, he insisted that new social conditions—especially industrialization and urbanization—widened the demand for expert knowledge.[13] Subsequent scholars discarded this functionalist framework, casting experts as clever operators who used jargon, intimidation, and outright fakery to prey upon lay fears and inadequacies.[14] Whatever their theoretical bent, however, all sides now presumed that expert authority diminished democratic rule: rather than serving the People, as Turner thought, experts had supplanted them.[15]

Over the past decade, two groups of scholars have amended this melancholy assessment of democracy and expertise in the early twentieth century. According to women's historians, reformers like Jane Addams and Florence Kelley developed an "alternative style of politics" to bring experts and laypeople into a smoother alignment. Self-proclaimed experts themselves, these women promoted professional authority and intervention with the same confident zeal as their male counterparts. Yet they tempered this hierarchical impulse with a commitment to democratic participation, born of necessity as much as philosophy. For women experts lacked all but the tiniest beachheads in the new administrative state: as originally constituted, for example, the female-inspired Children's Bureau boasted a staff of fifteen. To win "maternalist" measures like birth registration and mothers' pensions, then, female reformers had to enlist thousands of lay volunteers for publicity drives, door-to-door canvasses, and so on.[16]

Meanwhile, intellectual historians also undertook "a revision of the revisionist line" on expertise. Looking anew at John Dewey, especially, scholars argued that Dewey's theory of knowledge reflected—indeed, presupposed—a deep commitment to democracy. According to Dewey and his fellow "pragmatists," a term that also enjoyed something of a rehabilitation, human beings developed knowledge through a continuous, experimental engagement with the world. Open-ended and unpredictable, this process required tolerance, free inquiry, communication, and cooperation—in short, the skills and habits of democratic life. Insofar as experts tried to hoard knowledge, then, they impoverished both their own intellectual growth as well as the overall quality of democracy in America. "The essential need . . . is the improvement of the methods and conditions of debate, discussion, and persuasion," Dewey wrote. "That is *the* problem of the public."[17]

Yet most historical accounts focus upon experts' solutions to this problem, showing how this or that individual—Jane Addams, John Dewey, and so on—sought to reconcile democratic and professional authority. When we examine the problem from a lay perspective, as this book tries to do, we see that the reconcilers are right—but for a new set of reasons. Hardly innocent victims of expertise nor happy beneficiaries of its largesse, Mary Hunt and her WCTU assistants seized it actively and, most of all, judiciously: they took what they needed, as they needed it, and left the rest behind. *Laypeople* reconciled expert and popular movements, in other words, by pressing the former into the service of the latter.

A Clash of Cultures:
Liberals Versus Conservatives in the Progressive Era

In her preface to a 1936 collection of essays, Willa Cather pronounced a much-quoted requiem on the previous decade: "The world broke in two in 1922 or thereabouts. . . ."[18] Since then, a wide array of historians have confirmed her perception of the 1920s as a watershed in American cultural life. Nearly a half-century ago, Richard Hofstadter described a sharp split between "the village Protestant individualist culture" and "a new kind of educated and cosmopolitan American" following the First World War. The clearest example of the clash was the struggle over prohibition, pitting what Hofstadter termed "the immigrant drinking masses" and "the well-to-do classes" against a "rural-evangelical" coalition of ministers and farmers. Similar divisions appeared in countless other cultural debates, from sex in the movies to Darwinism in schools. Liberals stressed the social causes of human behavior, welcoming the erosion of a single, fixed standard of personal conduct. Conservatives rallied to defend this traditional framework, railing against the urbane elites who were undermining it. To scores of scholars now researching American conservatism, indeed, the 1920s mark not just the birth of liberal "modernism" but also of the modern Right.[19]

Yet the second half of Cather's comment hints at a somewhat different chronology. Although the break occurred around 1922, she wrote, ". . . the persons & prejudices recalled in these sketches slide back into yesterday's seven thousand years." America's cultural rift extended much further into the past, in other words, predating the 1920s by dozens (if not "thousands") of years. In Protestant churches, especially, deep divisions were apparent as early as the 1890s. Preaching a new "Social Gospel," liberal ministers demanded sewage treatment, housing inspection, and other municipal reforms that would reflect the loving spirit—if not the precise letter—of the Bible. Yet conservatives clung to a literal reading of Scripture, which warned that no true progress could occur until each individual found God. One side aimed to imbue the social order with the soul of Christianity, while the other imagined a new kingdom of converted souls.[20]

Before the 1920s, to be sure, these competing camps often galvanized around the same causes. Indeed, liberals and conservatives cooperated so readily that one contemporary observer detected a "Truce of God" between them.[21] Prohibition, in particular, enlisted not just the individualist Right (as Hofstadter thought) but the environmentalist Left.[22] Stressing the inherent iniquity of "demon rum," conservatives sought measures to interdict the sinners who imbibed it. Liberals, by contrast, demonized saloons and their wily operators for corrupting otherwise innocent patrons. Many of these liberal prohibitionists drank alcohol themselves, affecting an easy gentility that further distinguished them from their more austere right-wing foes.[23] Yet they also backed a ban on its sale, for the same reason they supported other reforms: it promised to improve Americans' worldly circumstances, which in turn would determine their individual fates.

At least two groups of critics dissented from this liberal creed in the years before World War One. Most famous are a small but influential set of elite "antimodernists," who worried that American culture was burying natural impulse and individuality in a bland social haze. Celebrating mysticism, medievalism, and even the Old South, these intellectuals hoped their spirited forays into the past would salvage personal authenticity in the future. Yet their crusade was essentially rearguard, a statement of what was desirable but not possible: like the revamped cult of the Confederacy, it derived much of its energy from intellectuals' perception that they, too, were fighting a lost cause. None of them pretended that they could restore the individual's old powers; instead, they aimed to protect a few meager shards of freedom and spontaneity from the grinding maw of modernity. Men like Henry Adams and William James were "individualists" only in the sense that "they wished to enhance individuality," as historian Thomas Haskell has noted. They were *not* individualists like nineteenth-century Americans, "who saw the social universe as actually composed of autonomous individuals who normally were masters of their own fates."[24]

Into our own century, however, millions of lesser-known Americans continued to view the world in precisely these terms. Displaying none of the pessimism that pervaded the antimodernists, Progressive-era conservatives insisted that individuals remained fully and morally accountable for their actions.[25] Yet they also embraced the growing trend toward expertise, belying many of the images we have inherited from historians of the 1920s. Following Hofstadter, most scholars depict early American conservatives as "localists" or even as "populists," stubbornly guarding their "traditions" against the encroachments of an impersonal, "modern" society—especially against the experts who managed it. Steeped in a venerable sociological idiom, such terms mask the ways that Americans could blend elements from each world.[26] For more than two decades, Mary Hunt fermented expert authority with individualist values. The result was Scientific Temperance Instruction, a deeply conservative wine in a distinctly modern bottle.

Textbooks Against Tipplers:
A Brief Snapshot of Scientific Temperance

"Would you like to know how to avoid this terrible power?" asked Charles H. Stowell's *Primer of Health,* after cataloguing the long list of health hazards caused by liquor. "Then refuse to take the first glass of cider, beer, wine, or any other drink that contains alcohol." Published on the eve of Hunt's death in 1906, Stowell's book was one of several dozen Scientific Temperance texts that she endorsed over a span of two decades.[27] Roughly half of the nation's school districts adopted one of these books to teach STI, which every state mandated by 1901. Since the vast majority of instruction occurred directly from textbooks—indeed, several states required STI texts "in the hands of pupils"—one of every two American schoolchildren probably received Mary Hunt's lessons in Scientific Temperance.

The books that she endorsed shared three major themes. The first pertained to the "nature," or "composition," of alcohol. Textbooks featured detailed descriptions of fermentation and distillation, tracing the development of beer, wine, and whiskey from barley, grapes, and corn. They also suggested classroom experiments with alcohol, such as pouring liquor over an egg or extracting it from yeast. These passages invariably concluded with the assertion that alcohol was a poison rather than a food, the sine qua non of Scientific Temperance. Throughout her life, in fact, Mary Hunt refused to approve any text that did not include that statement. "A poison is any substance which, when absorbed into the blood, will injure health or destroy life," declared a typical Hunt-endorsed book. "Alcohol is such a poison."[28]

As this remark indicated, though, the food/poison claim rested less upon the chemical origins of alcohol than upon liquor's physical effects. So the textbooks' second theme stressed the dangers of drink to an enormous range of human organs, from the brain, bones, and skin to the liver, lungs, and stomach. Even eyesight was imperiled. "Do you remember what we said about the red eyes of the hard drinker?" Stowell's *Primer* asked. "It is useless for such a person to ask the doctor to cure his eyes as long as he uses strong drink."[29] Intermediate and advanced texts offered more sophisticated explanations, often quoting experiments by "renowned" or "expert" European scientists. Rather than isolating this material in a single section, moreover, they usually spread it throughout the book. Each chapter—whether on respiration, digestion, exercise, or diet—concluded with a caution about the hazards of liquor to that particular system or function.

Finally, Scientific Temperance also emphasized alcohol's capacity for creating a "constant craving," or "diseased appetite"—that is, an addiction. "From the first glass of beer or wine sipped by the boy who is just beginning to drink to the dram of the drunkard whose tissues are poisoned by it," one book declared, "the nature of alcohol is to excite a thirst for more." Especially in texts for older students, this warning was frequently accompanied by an attack upon "light" or "moderate" tipplers. "No one can foretell how much resistance he may have against alcohol without taking a risk that may ruin his life," admonished another textbook, citing a "noted German professor" at Munich's Royal Institute of Hygiene. "To escape the mastery of alcohol, all drinks that contain it, even in small quantities, should be strictly avoided."[30] As Mary Hunt would discover, however, large numbers of distinguished American experts—scientists, physicians, educators, and even theologians—contested this total-abstinence claim. In the balance lay the fate of her movement—and, even more broadly, of lay influence upon American public schools.

Roots, Rivalries, Requiem:
The Rise and Fall of Scientific Temperance Instruction

The birth of Scientific Temperance bore little hint of its looming encounter with expert authority. Instead, as Chapter Two shows, STI clashed first with an older heritage of American localism. School boards condemned the new subject as a trespass into their areas of discretion, while school patrons complained that STI's teetotal message insulted their own drinking traditions. Galvanizing around Frances Willard, a longtime foe of Mary Hunt, state and district leaders in the Woman's Christian Temperance Union likewise charged that Hunt's national effort ignored their diverse creeds and conditions. To meet these grass-roots challenges, Hunt devised distinctly localist responses. Even as STI swept through state legislatures, her army of assistants converged upon their own schools and classrooms to enforce it. They also pressed Hunt's case within their respective WCTUs, belying critics' claims that she imposed her will upon these local unions. Like other nineteenth-century school controversies, in short, these early disputes hinged upon the proper *level* of democratic rule—state, local, or somewhere in between.

Starting around 1890, however, Hunt would collide with a new generation of experts that derided *any* democratic rule—at least in "technical" matters of school curricula. The first blows were struck by experimental physiologists, whom I describe in Chapter Three. Citing evidence from the research laboratory, their new professional lodestar, physiologists argued that STI exaggerated the documented dangers of alcohol. In response, Hunt enlisted her own scientific allies—from physiology, psychiatry, and clinical medicine—and funneled their claims to her assistants in the field. Streaming back into American schools, Hunt's expert testimony in hand, these lay volunteers insisted that the latest scientific research reinforced rather than refuted Scientific Temperance. Whereas both sides presumed that the laboratory would place STI beyond debate, then, its practical effect was precisely the opposite. At the highest level of "elite science," Mary Hunt managed to spark deliberation—and, as we shall see, recrimination—by recruiting her "own" handful of experts. As newspapers carried reports of this controversy to the countryside, meanwhile, Hunt's lieutenants argued loudly about "alcohol physiology" with school boards, principals, parents, teachers, and students.

Atop many of these school systems sat another breed of expert, professional superintendents, who gained increasing power over budgets, personnel, and curricula. Chapter Four details their attempts to unseat STI, which educators reviled as a stubborn lay intrusion upon their newfound authority. Embracing the "child-study" movement in psychology, these schoolmen also maintained that STI's scare tactics would harm fragile young minds. Here they joined hands with liberal theologians, whose Social Gospel eschewed the traditional Protestant emphasis upon individual sin and salvation. As in her struggle against scientists, Hunt first tried to draw upon her own experts to rebut this attack. Yet even sympathetic schoolmen were loath to link themselves publicly with a lay movement like STI, while most leading churchmen dismissed it as a vestige of their own "Puritan" past. So Hunt would have to draw on two other vocabularies, Scripture and democracy. Rejecting liberal readings of the Old and New Testaments, she noted that God himself—like STI—used fear as a method of persuasion; moreover, America's civic religion enshrined "the will of the people"—including, of course, STI—as sovereign. Despite the tension between these appeals, Hunt and her followers glided easily from one to the other. As the need arose, they insisted that Scientific Temperance reflected the sentiments of "the best men," the Bible, or the American citizenry at large.

Yet without the support of classroom instructors, Hunt also realized, no subject would ever reach American children. Since more and more teachers were female, especially in primary schools, she presumed that sisterly entreaties would win them over to STI. But teachers invoked their own professional prerogatives, Chapter Five shows, generally siding with male administrators against STI. Recent historical accounts of teacher professionalization presume that it was foisted upon them by these schoolmen, who used it to alienate teachers from "the community." On a question like STI, though, "the community" was starkly divided. Hence teachers' safest position was a retreat into the fortress of expertise, where they could fend off lay pressure of every stripe. Hunt, predictably, replied by copying them. As teacher resistance mounted, she sought to establish a graduate "College of Scientific Temperance," where a new set of STI instructors—or, in Hunt's rich phrase, "professional temperance workers"—would be trained. Once again, then, professionals and laypeople both embraced expertise in hopes of advancing their own agendas—and of escaping popular debate.

But debate continued, even within the Woman's Christian Temperance Union. After Frances Willard's death, a new set of national leaders started to mimic the liberal critique of Scientific Temperance. Citing the "the best men" in science and education, these WCTU officials blasted STI as both physiologically erroneous and psychologically dangerous. They also denounced Hunt's lay WCTU assistants for enforcing the subject in schools, echoing liberal claims that only trained specialists should execute public policy. For years, of course, Hunt had maintained that *her* expert allies were the "best men" and "foremost authorities" on the subject. So she would have to shift ground again, embracing the same fierce localism that had inflamed her initial opponents. Since her agents in the field best understood their respective school districts, Hunt argued, these women—not their liberal WCTU leaders—should determine the content and contours of Scientific Temperance. Here, again, she shuttled comfortably between professional and popular appeals. Trumpeting the titles and achievements of STI's expert supporters, Hunt simultaneously insisted that "the women," "the mothers," and "the people" supported it as well.

Upon Mary Hunt's death in 1906, however, this creative balance would melt away. Amid revelations that Hunt had collected royalties on the textbooks she endorsed, the WCTU disbanded her local army of lieutenants. Meanwhile, as Chapter Seven explains, a small core of Hunt loyalists refashioned her Boston headquarters into a clearinghouse of alcohol information. Without a grass-roots web of STI workers to transmit it, however, such research would never reach American classrooms. Instead, "school hygiene" stepped in to fill the void. Comprising vaccination and medical inspection as well as diet and exercise, this wider health curriculum perfectly embodied the growing liberal emphasis upon social environment rather than personal character. After the war, former Hunt secretary Cora Stoddard would briefly attempt to resuscitate her mentor's individualist ethos as well as her dense WCTU network of local volunteers. But these women quickly retreated from the task, lacking both the knowledge and the confidence that continuous, democratic debate—and debate alone—can bring. As the 1925 *Scopes* trial about evolution instruction seemed to demonstrate, Americans increasingly accepted or rejected expert dictates—in toto. When antievolutionist standard-bearer William Jennings Bryan declared that "it does not take an expert" to understand "the word of God," he served notice that the nation would now have to choose one or the other.[31]

In two ways, *Scopes* helps frame the end of this study. First, it remains our foremost historical symbol of what Hofstadter called the "unresolvable conflict" between popular democracy and modern expertise.[32] This book seeks to show that laypeople could and did resolve it, by harnessing expert authority to their own aims and purposes. Only after Mary Hunt's death would her lay assistants abandon this effort, inverting another common historical formulation. In most accounts of modern professionalism, experts forsake popular concerns for more narrow, technical ones.[33] In this book, by contrast, the laypeople desert the professionals. Just as lay initiative could reconcile popular and democratic authority, in other words, lay inertia paved the way for their separation.

The *Scopes* case also connotes "repression" and "intolerance," two charges that historians often level at Scientific Temperance.[34] Stubborn and single-minded, Mary Hunt indeed sought to silence all critics of her beloved STI.

Fortunately, though, her reach exceeded her grasp. As Scientific Temperance reminds us, even citizens who endeavor to restrict debate may—in practice— enhance it. For more than two decades, Mary Hunt and her lay assistants engaged Americans in a wide-ranging struggle over science, schooling, and democracy. This book seeks to chronicle that struggle and, most of all, to spark some debate of its own.

PART I
ROOTS: FIGHTING LOCALISM, 1879–1891

Men will vote for the education of the children before
they will vote for prohibition.
MARY H. HUNT, 1882

It is our duty not to take the word of some school official, but to visit
the school and carefully and wisely ascertain for ourselves if the
study is faithfully pursued by all pupils.
MARY H. HUNT, 1890

During its first decade, Scientific Temperance Instruction rolled through American state legislatures like a red-hot steam engine. At the helm sat Mary Hunt, who engineered victorious drives to require STI in thirty-three states between 1882 and 1890. But the subject stalled in the schools, where a venerable tradition of local school control blocked Hunt's dream of teetotal teaching across the land. Local leaders in Hunt's own Woman's Christian Temperance Union also resisted her directives, arguing that state and district officials should retain final discretion over the STI campaign. These foes rallied around WCTU president Frances Willard, whose bitter rivalry with Mary Hunt reflected a longtime rift within evangelical Christianity. A proud descendant of New England's early Puritan settlers, Hunt imbibed their predilection for reason and intellect over emotional outbursts. The Willard camp stressed passion as the fastest path to grace, insisting that any person could cast off sin—and find God—in a single, ecstatic stroke. Whereas Hunt called upon schools to carefully catalog the physiological consequences of drink, then, her WCTU foes promoted a more ebullient regimen of parades, pledges, and songs to dethrone "King Alcohol."

To fend off both local challenges, Mary Hunt forged a strong local response. When school patrons complained that Scientific Temperance insulted their community traditions, Hunt urged WCTU members from the same community to press her case with school boards, principals, and teachers.

Enlisting thousands of these women in a dense grass-roots network, mean-while, she also belied her critics' claims that she trampled upon local WCTU prerogatives. By the early 1890s, however, STI came under fire from a fresh generation of physiologists, educators, and theologians. Rejecting *any* form of lay influence in school affairs, these new expert enemies would provoke a new kind of rebuttal from Mary Hunt.

2

"A LITTLE BRIEF AUTHORITY": MARY H. HUNT, THE WOMAN'S CHRISTIAN TEMPERANCE UNION, AND THE BIRTH OF SCIENTIFIC TEMPERANCE INSTRUCTION

In April 1885, a triumphant Mary H. Hunt celebrated the passage of Pennsylvania's new law requiring lessons in "the nature and effects of alcoholic drinks." As the Woman's Christian Temperance Union's national superintendent for Scientific Temperance Instruction, she had traversed the state months earlier to spearhead its STI campaign. Five states already mandated the subject in their public schools; by the end of the year, nine others would do so. Yet Pennsylvania was special, Hunt maintained, owing to "its size and vast foreign population" as well as to the strength of its new statute. "It was the first law to which a definite penalty was attached," she wrote, "and hence an intense and varied opposition had to be encountered and overcome." From now on, schools that resisted STI risked losing their state moneys. Pennsylvania's children would soon be awash in Scientific Temperance, Hunt predicted, drowning any "wet" sentiment that still remained.[1]

A year later, however, reports from the field suggested otherwise. Carbon County certified that STI was "fully equal to that of the teaching of any of the other branches," exactly as the new law required. Yet in adjacent Monroe County, STI "was not very favorably received by the people, and a great many parents refused to supply their children with books," a local official noted. Dozens of other districts ignored the new subject altogether. Still others followed the letter of the law but not its intent, purchasing textbooks that endorsed—or at least acknowledged—moderate alcohol use. So WCTU women should pour into school boards and classrooms, Hunt urged, demanding that local authorities teach total abstinence. "The *enforcement* of the law is of vital importance," emphasized Josephine Weeks, Hunt's top Pennsylvania aide. "I would far rather we had *never* secured it, than that the children would be taught what is *not* true about alcohol."[2]

To Weeks's horror, even some local WCTU members supported so-called

15

moderate-drink textbooks. Here they followed Mary Hunt's arch-rival, national Union president Frances Willard, who praised these books in the WCTU's newspaper, the *Union Signal*. "With the National and State Superintendents saying one thing and 'our national organ' the very reverse, what a position we are placed in," Weeks moaned. "I believe it is part of the plan to destroy Mrs. Hunt's influence." Yet the text dispute concerned "partisan spirit" more than spirituous drinks, as another Pennsylvania observer noted.[3] The previous year, Hunt had broken with Willard over the latter's endorsement of the Prohibition Party. Willard proceeded to throw her considerable influence behind a competing WCTU educational department, the "Loyal Temperance Legion" (LTL). Both the LTL and the Prohibition Party featured parades, pledge signings, and other emotional displays, reflecting Willard's family roots in antebellum revivalism. Likewise, Hunt's support for Scientific Temperance—cool, rational, and nonpartisan—echoed the Puritan sensibilities of her own New England forebears, who resisted loud antics in favor of study and reflection.

In the end, every local chapter of the WCTU had to decide how—or even whether—to institute Scientific Temperance. Here they resembled local school districts, which still retained final discretion in most curricular matters. To rebut both challenges, Mary Hunt devised distinctly grass-roots responses. As STI swept through American legislatures, Hunt sent her assistants into their own school boards and classrooms to enforce these new measures. Within the WCTU, similarly, she appealed directly to local unions to initiate STI—and to eschew the directives of Willard and other rival leaders. Like most nineteenth-century struggles, these controversies hinged on the proper *level* of democratic rule: local, state, national, or somewhere in between. Only later would Hunt and her allies encounter an enemy that derided *any* democratic rule, at least in complicated school questions that were best left to experts.

A Puritan Heritage

Mary Hanchett Hunt was born in 1830 in South Canaan, Connecticut, to Ephraim and Nancy Hanchett. Ephraim, who operated a small family ironworks in nearby Salisbury, was of nebulous Welsh extraction; Nancy descended directly from Edward Winslow, an early governor of Plymouth colony, and Thomas Thatcher, the first pastor of Boston's Old South Church. Throughout her life, Hunt and her admirers would boast of her mother's glorious "Puritan stock"; the "sturdy miners" of her father's lineage drew fewer accolades.[4]

Hunt was equally proud of her birthplace in Litchfield County, home to the legendary minister and temperance leader Lyman Beecher. Beecher founded Litchfield's first temperance organization and later helped spearhead the American Temperance Society, which dominated national antidrink efforts into the 1850s. As a theologian, meanwhile, he was best known for his attacks upon the boisterous evangelical revivals that swept across antebellum America. Discarding the grim determinism of his own Puritan forebears, Beecher conceded that humans—properly instructed—could help will their salvation. But he ridiculed the perfectionist zeal of preachers like Charles G. Finney, whose noisy prayer meetings promised instant conversion to anyone who chose it. Such rituals violated Beecher's New England decorum as well as his ongoing sense that man was "rebellious, sinful, and odious to his Maker." Contemplation and reflection, not passion and delirium, paved the road to Heaven; and some sinners were simply too hardened to get there.[5]

The same cautious optimism marked Beecher's thirty-year struggle against alcohol. Again, for Beecher and his colleagues, reason held the key: by publicizing the physiological, social, and spiritual consequences of drink, they would persuade right-thinking Americans to cast it out. But this approach came under challenge in the 1840s from the Washingtonians, bands of "reformed drunkards" whose loud gatherings echoed the cadences of evangelical revivalism. Like Finneyite ministers, Washingtonians centered their meetings around testimonials, or "experience sharing"; and they too held out the promise of immediate redemption, claiming that any tippler could be "cured" in a single ecstatic stroke. Some Washingtonians even went on to deny "the basic depravity of man," as one angry critic noted. Herein lay the last real heresy, in the eyes of New England's neo-Puritans. For despite their new accent on human redemption, they still insisted that certain men lay beyond it. "Keep the temperate people temperate," implored Justin Edwards, a close Beecher associate. "[T]he drunkards will soon die, and the land be free."[6]

Sadly, our sparse accounts of Mary Hunt's adolescence do not record her reactions to these affairs. Her father quickly came under Lyman Beecher's sway, serving as a vice president of the Litchfield Temperance Society after Beecher established it. Both of her parents abstained from alcohol, even refusing to serve it at so typically a besotted an event as the raising of a new forge frame. (Instead, a Litchfield newspaper later reported, Ephraim Hanchett's tired assistants received "a sumptuous hot dinner of every possible viand"—courtesy of Nancy Hanchett, "the young Puritan wife.") The Hanchetts' convictions made an early and powerful impression upon Mary, as an embarrassed young suitor would discover. After attending a wedding, he and the fourteen-year-old Mary had accompanied the betrothed on the beginning of their honeymoon. Along with several other couples in the party, they spent the first night in a hotel. In the morning, Mary came downstairs to find her escort at the bar, "tossing off" a shot of liquor. "That settled the matter," wrote a Hunt aide; "he waited upon her no more. Soon after this she went away to school."[7]

In 1848, Mary entered the Patapsco Female Institute in Ellicott City, Maryland. Here she came under the tutelage of Patapsco's dynamic principal, Almira Hart Lincoln Phelps, whose fears of emotional excess and of a stern Deity closely followed Lyman Beecher. Like her sister and fellow educator Emma Willard, Phelps was an adult convert to Episcopalianism. Yet she continued to worship the angry God of her own Puritan ancestors, insisting on the essential depravity of humans before him.[8] In the same vein, Phelps also prohibited any expression of religious exuberance by Mary Hanchett and her other pupils at Patapsco. "I would address you as rational beings," Phelps told the students during Mary's first year at the school. Revivalist histrionics interfered with the sober reflection that distinguished sophisticated Christians, Phelps maintained. A quarter-century later, in fact, she would condemn the Woman's Christian Temperance Union for promoting "unhealthy excitement" in the name of true temperance.[9]

Hunt and the WCTU: From Gospel to Scientific Temperance

Mary Hanchett taught briefly at Patapsco upon her graduation in 1851, then married ax manufacturer Leander B. Hunt. His company boomed, but he was forced to withdraw in 1865 when a "dishonest partner" stole the profits and "some investments failed to materialize as Mr. Hunt had expected," as an associate of Mary delicately noted. The couple stayed afloat by taking boarders at their newly purchased home in Hyde Park, Massachusetts, one of Boston's first "streetcar suburbs," where their facade ornaments and manicured lawn marked the Hunts as solidly middle class. So did Mary's voluntary activities outside the house, ranging from Bible study and Sunday school teaching to a brief stint as president of Hyde Park's "Ladies' Sewing Society."[10]

It is unclear just when Mary Hunt joined the Woman's Christian Temperance Union. The WCTU arose in the dramatic "woman's crusade" of 1873 and 1874, a spontaneous wave of protests against saloons; after that, it sponsored prayer meetings and other "spectacular" public events. Elected vice president of the fledgling Massachusetts WCTU, Hunt initially recruited drinkers to sign a pledge of sobriety. Like Almira Phelps, though, Hunt quickly soured on this Gospel Temperance approach. After an especially fruitless effort among alcoholic prisoners, she concluded that moral appeals could never penetrate such wayward souls. "A few of them go over to the temperance side, but fewer of them stay," Hunt complained, echoing Beecher and other early critics of such efforts. "The majority slump back. The world's hope in their case is in their death."[11]

So Hunt directed her energy towards prohibition, joining Frances Willard and other young WCTU leaders who also wished to broaden the organization beyond its Gospel Temperance base. In 1879, she appeared before the state legislature's license committee to press for an outright ban on liquor. Here the future champion of Scientific Temperance Instruction mocked the old suasionist bromide that only "education" could limit drink. "He whose laws control the Universe did not wait until men were educated up and away from selfishness, before He gave the law," Hunt told the committee. Rather, she argued, God gave laws first—and "the law is a school master," as Scripture says. Nor could Hunt countenance the standard objection that prohibition would provoke ridicule and resistance. "Not enact a righteous law, because some men will disobey it?" she asked. "O fallacy, fallacy!" We might just as well repeal our laws against theft, Hunt reasoned, on the grounds that citizens steal.[12]

Just as Hunt had rejected moral solutions, however, she soon began to question legal ones. Prohibition went down to defeat at the statehouse and in dozens of localities, revealing a deeper public antipathy than Hunt had anticipated.[13] Yet the turning point came when she petitioned the legislature to ban public smoking, another common WCTU target. Such a law had been on the books for years, Hunt discovered, but was a "dead letter" in practice; recognizing as much, state lawmakers responded by repealing it. "I learned from that incident that the people's sanction of a law is found in its enforcement," Hunt recalled. "If that sanction is wanting the law looses [sic] its cogency." Prohibition remained the final goal, of course. Yet it awaited approval from the "jury"—not "twelve men chosen from the masses," Hunt declared, but "a *majority* of the 54 millions who are our countrymen."[14]

To forge this majority, Hunt looked to America's common schools. Since the 1830s, schools had sought to inculcate "temperance principles." But here, too, gospel-type techniques predominated. Children marched in "cold-water armies," declaring independence from "King Alcohol"; less often did they study the oppressive effects of liquor upon the stomach, liver, and other organs. Even after the Civil War, school temperance campaigns continued to stress parades and pledges over pathology and physiology. For Mary Hunt, however, such efforts violated the canons of reason and reflection at the heart of her Puritan heritage. Following Phelps, who authored popular textbooks in botany and other sciences, Hunt presumed that "all men desire knowledge." To be sure, some people—drinkers, especially—had forsaken this faculty. But children still possessed it. Denouncing cold-water appeals, then, Hunt argued that temperance lessons should reflect "cold logic and scientific demonstration" instead.[15]

Hunt's own interest in the "scientific" approach was piqued during an experiment by her son Alfred, a student at the Massachusetts Institute of Technology. Assisting him with the experiment, Hunt "thought there was something wrong with the alcohol he was using." Her curiosity led her to the pages of *Popular Science Monthly* and a rich heritage of alcohol research, dating all the way back to the eighteenth-century physician and patriot Benjamin Rush. Like Lyman Beecher, who borrowed liberally from him, Rush ranked and cataloged the myriad effects of "strong"—that is, distilled—drink. More recently, Hunt found, another generation of investigators had sought to demonstrate the dangers of all liquors, fermented and distilled alike. The key figure in this movement was British physiologist B. W. Richardson, who declared that alcohol in any form or amount was poison. Such research promised to boost total abstinence beyond mere "moralistic declarations," which no longer sufficed in a new era of science. Cool and rational, optimistic but not utopian, this appeal fit perfectly with Mary Hunt's Puritan temperament.[16]

Combining Two Suasions

Armed with Richardson's *Lectures,* Hunt visited her local school committee to ask that children receive lessons from it. The committee complied, adopting several "scientific temperance" textbooks for Hyde Park—and catapulting Mary Hunt into a much wider spotlight. Hunt's victory caught the eye of Frances Willard, who was herself about to win the presidency of the Woman's Christian Temperance Union. At Willard's invitation, Hunt addressed the WCTU's 1879 national convention in Indianapolis. The convention anointed Hunt director of its new "Temperance Text-Book Committee," soon to become the Department of Scientific Temperance Instruction. Hunt then called on all WCTUs to select state, county, and local "superintendents" who would visit school boards to demand STI, just as she had done. By 1880, Hunt would gleefully report, boards were "in a state of siege at the hands of the mothers." Dozens of towns had already adopted temperance texts, buckling under this barrage of matronly might. Others were sure to follow.[17]

Yet Hunt's hopes proved premature, forcing a shift in strategy. Just a year later, the WCTU convention would resolve to seek temperance education laws in state legislatures. School committeemen, it seems, were not as pliant as Hunt had pictured. "[W]e have found our Gibraltar in the district board," complained a Michigan woman. "These gentlemen are respectful to the voting and tax-paying power that elects them . . . and in them a solid wall has been found, especially in the cities." Too often, boards of education were "elected by a drinking constituency"; predictably enough, many committee members were tipplers themselves. "Beer and whisky towns and cities have beer and whisky school boards," Hunt noted, "who call total abstinence teaching 'fanatical.'" These boards, she realized, "could not legislate in opposition to their views." Instead, the state must legislate for them.[18]

Such a claim, of course, contrasted sharply with Hunt's critique of prohibition. Since "a government of the people cannot compel majorities," she wrote, voters "must first be convinced that alcohol and kindred narcotics are by nature outlaws, before they will outlaw them." The same logic would suggest that school boards must first be convinced that STI was sound, before they would submit to it. Yet Mary Hunt's legislative strategy was conceived in the belief that some boards could *never* be convinced. In this sense, Scientific Temperance bridged two mainstreams of nineteenth-century temperance—indeed, of nineteenth-century reform. Simultaneously coercive and persuasive, STI laws would use legal suasion of adults to institutionalize moral suasion of the young.[19]

Legislating Scientific Temperance

Hunt began this new mission in neighboring Connecticut, in a confident mood. "Plead [*sic*] for the State Mandatory Law for Educational Temperance," she wrote in her diary on 3 February 1881. "God will take care of results." Yet lawmakers in the Nutmeg State rebuffed STI, refusing even to report out the bill. Clearly, then, legislative victory required preliminary legwork. Traveling north to Vermont the following fall, Hunt organized a massive petition campaign for STI. She eventually addressed the legislature herself, but only after each member had heard from his constituents. "The people are the real source of power—they must be the lobby," Hunt wrote. On 13 November 1882, Vermont became the first state to require Scientific Temperance in all of its public schools.[20]

But it was also a traditional temperance stronghold, as Hunt's allies admitted.[21] The real test of Hunt's new strategy would come in Michigan, where STI faced severe resistance. Arriving just after the Vermont victory, Hunt quickly set about lecturing near *the residences of Representatives and Senators elected members of the nearcoming legislature* [italics in original], as an admirer reported. With "womanly tact," local WCTU members "secured the presence of these soon to be lawmakers" at Hunt's speeches. They also obtained signatures from "influential constituents" in each man's district, sending these petitions on to Lansing as soon as the new legislators arrived there. Lest veteran lawmakers miss the point, women presented the petitions every morning on the floors of both houses—from the first day of the session right up until the final vote.

Meanwhile, WCTUs sent special entreaties to the president of the Senate and speaker of the House, asking them "to remember the wishes of the people on this subject" when they appointed the joint committee on education. ("[A] favorable committee was thus secured in the Legislature," a Michigan woman reported.) The committee quickly granted a hearing to Hunt, who also addressed a special combined session of the House and Senate. Each member of the legislature received personal invitations to this speech, not just from the state WCTU but also from "the various local W.C.T.U.'s in his Representative or Senatorial District." All the while, WCTU women kept STI "apart from all other petitions" so "it might come before the people and Legislatures an entirely non-partisan movement." When the bill passed, all those voting in favor obtained a "card of thanks" and a bouquet—courtesy, of course, of the Woman's Christian Temperance Union.[22]

Thereafter, this "Michigan Plan" became the centerpiece of Mary Hunt's national strategy. Twenty-one state legislatures would mandate Scientific Temperance during the next five years, wilting under the WCTU's steady hail of letters and petitions.[23] In Congress, meanwhile, Hunt would spearhead a successful drive to require STI in all District of Columbia and territorial schools. Taking up residence in Washington for five months, she identified critical lawmakers and solicited seventy thousand signatures from their constituents. Even STI's occasional setbacks demonstrated the power of "systematic platform work *among the people*," as Hunt emphasized. In Ohio, for example, a train accident disabled Hunt and derailed the campaign shortly before a key vote on an STI bill. Hence "petitions were presented to the legislature with the essential preliminary work . . . only partially done"—and the measure went down to defeat.[24]

STI and Partisanship

As Hunt often noted, STI also benefited from the sharp partisan nature of other nineteenth-century disputes over alcohol. Nothing polarized America's political parties more than prohibition, which pitted "dry" Republicans against "wet" Democrats in state after state. Predictably, then, STI found its staunchest supporters among the GOP. Critics were almost always Democrats, who quickly identified STI as an easy target for satire. A New York lawmaker offered an amendment that would require each school district to place a "fearful example"—that is, a drunkard—at all STI examinations; an Iowan sought to insert a provision for teaching the effects of the tariff, especially "upon the human system of the poor and laboring masses"; and a Pennsylvania legislator, provoking "a roar of laughter," wondered how Washington and other Revolutionary leaders had performed such heroic deeds without STI. A fourth Democrat set his own satire to rhyme:

> But now, Anno Domini, eight-five
> Blessed woman evolves a plan
> By which she eventually hopes to secure
> The absolute control of man
> On his wine and beer, tobacco and cigars
> She will place a permanent lien,
> By teaching the boys and girls to shout
> Physiology and hygiene.[25]

But in the end, a Connecticut observer noted, these same critics often cast their votes for Scientific Temperance. Satirical jibes "provoked a great deal of amusement," he remarked, "and everybody at once saw the point of the shaft, although they did not feel [it] severely enough to vote against the bill." Elsewhere, even in notoriously wet states, large numbers of Democrats lined up behind STI. In Pennsylvania, for example, twenty-five Democratic representatives backed STI and twenty-seven opposed it; with the GOP voting almost unanimously in favor, it passed easily. Likewise, Democrats in New York and Illinois split almost evenly. The most astonishing tally came in New Jersey, where a well-known saloonkeeper led the charge for Scientific Temperance—and every Democrat supported it. "Men," Mary Hunt declared, "will vote for the education of the children before they will vote for prohibition."[26]

Indeed, Democrats often rebutted prohibition with their own calls for "education." Rather than banning alcohol, the argument went, society must persuade Americans to avoid it. Of course, this was precisely the argument for Scientific Temperance. Hence even George Washington Glick, colorful leader of Kansas's wet Democrats, supported STI during his successful 1885 gubernatorial campaign against prohibitionist John St. John. For two decades, Glick had vouched for the virtues of "moral suasion, of the old Washingtonian type," as he liked to say. He could hardly oppose STI, which seemed—at least to Glick—to embody these very virtues. Nor could Republican drys object, since it also contained an element of legal suasion. Indeed, a Californian observed, Scientific Temperance was

> a measure so reasonable, so just and sensible, so free from all entanglements with controverted and controversial subjects that all alike, conservative and radical, prohibitionist and anti-prohibitionist, woman suffragist and anti-woman suffragist, democrat and republican and everybody can harmoniously unite.[27]

Yet early reports from the schools suggested otherwise. Three years after STI sailed through Michigan's legislature, for example, only a third of its school districts were teaching the new subject. There was "a very general disposition on the part of district officers to either ignore the provisions of the law, or to deliberately disregard them," as the state school superintendent complained. Out in the schools, it seemed, legal suasion ended where localism began.[28]

Local Resistance I: The Schools

From its very inception, America's common school movement clashed with a tenacious tradition of local control. Men like Horace Mann contemplated—but never consummated—vast state systems of public education, which sprouted haltingly across the antebellum North. School bureaucracies boomed after the Civil War, as most states enacted provisions for county superintendents, compulsory attendance, and so on. Yet this growth—like its modest antebellum antecedents—proceeded in staccato rhythm, with large-scale buildups punctuated by occasional contractions. Moreover, the new systems failed to achieve the authority that Mann had craved. Localities continued to ignore school statutes and to mock state schoolmen, whose angry pleas only accentuated their impotence.[29]

Most of these officials seem to have welcomed Scientific Temperance, since it echoed their own centralizing impulse.[30] By the same token, however, they usually proved unable to execute it. "Patrons refuse to have this branch taught," wrote an angry Iowa superintendent, "and no power is enforced to bring about the desired end." Part of the problem stemmed from the nebulous language of STI laws, which typically required that "all pupils" receive "regular" instruction in the new subject. "[I]t cannot be expected that a law so loosely worded can effect much in a short time," groaned a Delaware school official, "as the instruction may be given only once a week or even once a month." The result was a crazy quilt of requirements, reflecting "the personal habits and prejudices of the school board and patrons," a third observer noted. San Francisco provided almost twice as much STI as Los Angeles; Grand Rapids furnished three times as much as Detroit; Omaha reserved part of its weekly "language lesson" for the subject; and at least one North Dakota county taught STI every day. New York City, by contrast, declared that a single monthly lecture of twenty minutes would suffice. Rochester required just one STI lesson per term, while Syracuse ignored the subject altogether.[31]

The sharpest resistance often came from immigrants. Eager to rebut charges of indolence and drunkenness, foreign-born state legislators rallied almost unanimously behind Scientific Temperance measures.[32] Yet given the ambiguity of these laws, which generally required instruction in "the effects of alcohol," local districts could depict strong drink in whatever manner they chose. Frequently, immigrants charged, the resulting curriculum insulted both their cultural traditions and their parental prerogatives. "My boy tells me that when I drink beer der overrcoat vrom my stummack gets to thick," complained an 1896 "note to teacher" in Mary Hunt's *School Physiology Journal*. "Please be so kind and don't intervene in my family affairs." Patently apocryphal, Hunt's anecdote nevertheless contained a note of truth. The following year, a coalition of German societies in Chicago complained that STI "abet[s] the views of the Prohibitionists . . . under cover of Physiology." Elsewhere, too, immigrants began to blast STI as "prohibition agitation." In statehouses, of course, they had welcomed Scientific Temperance for *impeding* prohibition: instruction about alcohol would suppress its use, the theory went, obviating any need for bans on its sale. Once STI reached the schools, though, immigrants often came to regard it less as a bulwark against dry laws than as a harbinger of them.[33]

Still, as Hunt admitted, the vast majority of STI's local opponents were native-born Protestants. Some foes simply objected to the added cost of the new subject, as a Maryland school principal observed. Others condemned it for the same reason immigrants did: it violated their own drink habits.

Borrowing a standard piece of prohibitionist boilerplate, Hunt frequently blasted besotted factory laborers and urban elites—"the working classes and the wealthy classes"—for resisting Scientific Temperance. Yet wine and other liquors were enjoyed by growing fractions of the middle class, too, even by so-called respectable women. Hence STI's enemies defied simple categorization by ethnicity, class, region, or even gender, reflecting what one WCTU activist called the "different drinking customs" of the nation itself. From New York to Michigan to Washington, parents kept their offspring home to avoid Scientific Temperance. When an Iowa school taught it at "an unexpected time," meanwhile, critics fumed that "the children were taken advantage of, and deceived into hearing what their parents disapproved."[34]

Hunt's Response: "Study Your Men"

At Hunt's urging, meanwhile, WCTU women flooded back into American schools to block the localist challenge. "School examiners, school boards and school superintendents are, most of them, indifferent to the law—ignore it—and are not dismissed," complained an Ohio woman. But "no law will enforce itself," she added. Fulfillment would demand the renewed vigilance of WCTU women, who continued to cajole stubborn local officers. "It is amusing when girls . . . go to the superintendents of schools," wrote a Virginia woman, "and when these worthies will not come to terms, to hear these girls say, 'But no matter what you think, the code of Virginia requires you to teach it.'" To prepare local STI workers for these negotiations, WCTUs across the country sponsored "mock school boards" where an STI "visitor" (or Hunt, if she could attend) fielded objections from women masquerading as irate board members. Additional counsel came from Hunt's constant barrage of pamphlets, circulars, and personal letters to women in the field. To sway a school official, "do not go prancing up to his gates on your hobby horse, aggressively demanding admittance," one such missive warned. "Remember also the exceeding sensitiveness [sic] to their own dignity which comes with a 'little brief authority.' The average school director is enveloped in it. . . . You must study your men. What will have great weight with one will have little with another."[35]

Most importantly, Hunt maintained, local STI workers must visit classrooms to observe lessons, examinations, recitations, and textbooks. "It is our duty not to take the word of some school official," Hunt wrote, "but to visit the school and carefully and wisely ascertain for ourselves if the study is faithfully pursued by all pupils." In each of the twenty-three towns in Massachusetts's Hampshire County, for example, a "superintendent" selected by the local WCTU monitored the schools and reported to county STI superintendent H. A. Orcutt. Orcutt's 1889 records reveal that these local agents collected information on 292 schools, noting for each town the total number and grade level of students receiving STI, the nature of examinations, and the type of textbook (if any) in use. In regions too remote to support a WCTU, the state union often recruited a local woman to inspect the schools. Thirty such agents served in Maine, yielding a state total of 128 local superintendents even though Maine had only 98 WCTUs. By 1898, Pennsylvania would boast 774 superintendents, the largest statewide total; Wisconsin won the distinction for the most new agents enlisted in that year, 34.[36]

Lest each of these workers pursue a separate agenda, meanwhile, Mary Hunt also prepared and distributed a "Standard of Enforcement" for Scientific Temperance. Signed by several prominent state educational leaders, the standard called for at least three STI lessons per week for fourteen weeks a year and physiology textbooks "in the hands of pupils." Furthermore, it specified that books should devote at least one-quarter of their space (or, at the high school level, twenty pages) to alcohol and narcotics. Later, in the face of continued local resistance, Hunt would press legislatures to amend their STI laws so that these elements became compulsory. By 1895, for example, sixteen states mandated textbooks for students who could read. Even better, Hunt boasted, most statutes added a penalty—in either fines or withheld state moneys—for districts that defied them. From bitter experience, Hunt wrote, she had learned that "a law without penalty is only advice."[37]

Yet even these allegedly airtight measures resisted "exact interpretation," as a Colorado official complained. His own law required textbooks "in the hands of pupils," but did not specify whether state educational departments, local school districts, or parents would have to purchase them. Penalty provisions were even murkier. Although Iowa's statute denied state funds to offending districts, for example, schools continued to snub the law because it failed to describe "just how forfeiture is to be worked," a critic noted. Where fines were mandatory, meanwhile, it was often unclear whether they should fall upon guilty school officials or upon a district at large. Even after states added penalty provisions, then, local resistance continued. In Michigan, for example, one-half of school districts still refused to provide STI—and none of them suffered a loss of revenue, Hunt's new amendment notwithstanding. Given the weakened spring of state authority, in other words, even a law *with* penalty was "only advice." So Mary Hunt would augment it, via her flexible legions of local assistants.[38]

The Textbook War

The most vehement battles surrounded textbooks, which were usually selected at the district level. "That we may teach is settled," Hunt declared in
1887, after all but four northern states had mandated Scientific Temperance.
"What shall we teach? is now the question to be widely answered." Hunt
never hesitated in her response: all textbooks should declare that alcohol is
poison, potentially lethal in any amount. Hence she urged local STI workers
to win adoption of teetotal texts like A. S. Barnes's Pathfinder series, which
bore Hunt's personal endorsement on its inside cover. Yet especially in the
West, the fastest-growing textbook market, school boards often selected
books with a more moderate message. After a German school board member
condemned the Pathfinders, for example, St. Louis adopted an "abridged"
version that described alcohol not as a poison but simply as "a clear liquid,
about the color of water." Gone also were the Pathfinders' earlier statements
that alcohol "hurts both body and mind" and "changes the person who
drinks it in any form."[39]

Beyond the Gateway, hundreds of other communities selected Barnes's
"abridged" Pathfinder or even William Thayer Smith's *Physiology*, which declared flatly that "a little liquor" makes men "feel comfortable or gay." "[W]e
cannot work successfully in this whiskey ridden, wine producing state a book
which . . . makes the Temperance question the main thing," wrote Barnes
agent Edward F. Adams from California, with reference to the original Pathfinders. "It may be glorious to fight for it, but I think the result would
generally be the adoption of Smith's." In Nevada, for example, WCTU
women "did all they could" for Hunt's endorsed books—yet the state board
of education, which selected textbooks in the Silver State, unanimously approved Smith's. So did school boards in Butte and dozens of other Montana
towns. Only in territorial Utah, ironically, did boards rally around the original Pathfinders. "Utah is the boss! The Mormons forever for temperance!"
Adams exulted. "They say this is the one law which Congress has passed
which they believe in, and they want to show their loyalty on this." (To
illustrate the point, Adams enclosed a letter "from a live Mormon"; clearly
enjoying himself, he warned all readers to "hold it in a pair of tongs.")[40]

In the West, Smith's and the Pathfinders ran a two-horse race; back East,
by contrast, dozens of companies charged after the STI book trade. Pennsylvania's 1885 law released a veritable waterfall of book salesmen, pouring
into the state to promote their wares via the usual mix of cajolery and chica-
nery. Some agents simply offered bribes; others showered the state with free

samples, the fastest way to a parsimonious school board's heart. As district after district rejected her favored Pathfinders, then, Hunt accused rival companies of conspiring with crooked local officials against "the motherhood" of Pennsylvania. "Are not the text-book committees the people's servants," she asked, "and are not the women people?" In Philadelphia, an aide added, 150,000 women petitioned the school board on behalf of Barnes's series. Yet "the individual wishes of book agents are considered superior to theirs," she lamented, so a rival text was adopted.[41]

Ironically, Hunt would also condemn Barnes's own "book agents" for failing to promote the Pathfinders *enough*. In response, Barnes's top salesman charged that WCTU women had prejudiced school boards against the books. They were both right. Barnes pushed its abridged edition more avidly than the Pathfinders, eventually publishing a third edition that was even weaker in its condemnations of alcohol. Across the state, meanwhile, WCTU visitors offended local school officials from the "liquor and beer classes," as a rival book agent confirmed. Here he alluded to Pennsylvania's large foreign-born population, which took special umbrage at the Pathfinders' teetotal provisos. Simultaneously, though, Hunt's efforts were also under assault from a much less predictable quarter: her own Woman's Christian Temperance Union.[42]

Local Resistance II: The WCTU

"Our women in Iowa have always felt that you assumed dictatorship too much. I have told you this often." In December 1885, Iowa WCTU president J. Ellen Foster told Mary Hunt again: Iowans resented her efforts to control their campaign for compulsory Scientific Temperance Instruction. "Mrs. Hunt said . . . she could go to no state unless she *managed* the whole business," one WCTU critic wrote, in a letter to Foster, "and Iowa women are not the kind of women who can submit to be managed." True, the critic acknowledged, Hunt's lobbying had already produced several dramatic victories back East. Yet "Iowa is not Pennsylvania or New York," she emphasized; its WCTU women "know Iowa better than Mrs. Hunt." They would never "yield . . . to a foreigner," she added, "even though she is National Superintendent." Citing several more complaints, Foster stressed that her constituents were "*intensely Iowa* in their sympathies." Lest Hunt spark even more hostility, Foster urged her to stay away from the Hawkeye State—and to cede STI to Iowa's own Woman's Christian Temperance Union.[43]

In Mary Hunt's view, such critics "had not yet come to understand the relation of each worker to the work." Within the Woman's Christian Temper-

ance Union, however, this relation was rarely as clear as she implied. Originally composed of state, district, and county unions, the WCTU grafted
twenty-one functional departments—including Scientific Temperance Instruction—onto its divisional structure in 1880. (Mimicking "the political subdi-
visions that men have found convenient," county unions would add local
chapters in 1884.) Theoretically, unions at each level retained the right to
choose which departments they would adopt; likewise, functional superintendents were to be selected by the auxiliary unions themselves, not by departmental leaders. In practice, however, departments often operated like independent fiefdoms. Assuming the national STI helm in 1880, Mary Hunt soon
boasted "[a] great army with battalions in every State and territory, regiments
and companies in all counties, with pickets at every outpost, and all under the
guidance of one head": herself. To assure their loyalty, Hunt often recruited
state STI superintendents on her own; these agents, in turn, chose lieutenants
at district and county levels.[44]

From the very start, however, state WCTU officials argued that *they* should
command Scientific Temperance. "Vermont ladies feel that it was scarcely fair
. . . to give [Hunt] *all* the credit for carrying the school law in the State,"
wrote one critic, following passage of the nation's first STI measure in 1882.
"They think their own leaders were entitled to fully half the credit." The next
year, Massachusetts WCTU president Mary Livermore refused Hunt's request to address the state's annual convention. Despite protests by STI superintendent Emily Laughlin, a staunch Hunt ally, Livermore argued that "local
ladies" could handle the subject without "outside" assistance. Yet Hunt herself
hailed from Massachusetts, as an enraged Laughlin emphasized. "[R]emem-
ber I thought Mary Livermore was a Christian," Laughlin wrote. "I hope
God will turn and overturn in our dear Old State 'till the devices of the
wicked shall be brought to naught."[45]

Similar tensions surrounded the subject in New York, where superintendent Elizabeth Greenwood distributed Hunt's 1883 "plan of work" for winning STI laws. New York WCTU president Mary Burt charged that the
plan "ignored the State," since no local union officers were consulted before
its circulation. Worst of all, Burt complained, the plan identified Hunt secretary Cornelia Alford as *New York's* STI secretary, even though state leaders
never approved Alford's appointment. Beneath Burt's remark lay "the determination . . . to keep [Hunt] from coming into the State to work," Alford
warned. An unfazed Mary Hunt proceeded to deliver no fewer than fifty-
seven speeches across New York, braving heavy snows as well as Burt's
hostility. In accord with her plan, local WCTU women barraged their legis-

lators with petitions, leaflets, and letters. New York passed an STI measure in 1884, the "crowning victory" in "our four years' struggle," as Hunt declared. Simultaneously, however, rivals like Burt and Livermore were galvanizing around Hunt's strongest foe of all: Frances Willard.[46]

Willard and Hunt I: The Partisan Background

During STI's first few years, Willard and Mary Hunt seem to have cooperated smoothly. The break came in 1884, just after Hunt's New York triumph, when Willard secured the WCTU's endorsement of the Prohibition Party. To Mary Hunt, this policy made her own work "almost impossible." A lifelong Republican, she fully recognized that WCTU women were divided along party lines. But that was all the more reason to maintain the facade of neutrality, Hunt insisted. For if the WCTU became linked to a single party, she maintained, Scientific Temperance would suffer the same grim fate as prohibition. Arriving at the WCTU's 1885 convention, Hunt resolved to press for a return to its older, "non-partisan" policy.[47]

In the interim, however, no fewer than nine more states passed mandatory STI laws. "They told us they could not . . . secure laws because of the partisan attitude of the W.C.T.U.," gloated one Willard ally. "Now what are the results of this year's work? We have never had such favorable legislation." Actually, Hunt countered, seven of these measures lacked the penalty provisions she had sought. A "partisan" WCTU could only win a "law without penalty," she concluded. But the convention was not persuaded. Only 29 of 275 WCTU women joined Mary Hunt in condemning partisan endorsements. The rest of the delegates—including Mary Burt and 26 other state presidents—stood solidly behind Frances Willard.[48]

Meanwhile, Willard and Hunt tangled on another "endorsement" issue: STI textbooks. Shortly after Hunt approved A. S. Barnes's Pathfinder series in 1884, Willard announced her own preference for a competing text, Orestes Brand's *Lessons on the Human Body*. "I have all along told [Barnes] that I, as National Superintendent, represented the interests of this phase of our work," Hunt grumbled to Willard. "You see this puts me in a bad light with my publishers, unchurches me. You will right me for the sake of the work, I know." Willard instead went on to approve the even more controversial "Smith's" text, which also omitted the claim that liquor was poison. Her decision might have reflected Willard's more tolerant attitude towards alcohol, which she used medicinally. More likely, though, this second endorsement dispute simply reflected animosities from the first. "In my opinion this is all from the Third Party movement," one observer told Hunt, after conferring with Willard about the textbooks. "You are on the wrong side and must expect sharp criticism on your work."[49]

Willard and Hunt II: The Religious Background

Willard and her allies also shared an emotional religious style, diametrically opposed to Hunt's Puritan sensibilities. Like Mary Hunt, Willard came of age during the feverish evangelical revivals that swept the antebellum North. But Hunt's parents were neighbors and allies of Lyman Beecher, who maligned the revivals' emotional excesses and overblown rhetoric. The Willard family embraced Beecher's main rival, Charles G. Finney, whose fiery sermons promised immediate redemption to anyone who sought it. Willard was born in the "burned-over" district of western New York, site of Finney's most famous campaigns. When she was two, the Willards set out for Ohio to join Finney at his newly created Oberlin Collegiate Institute. For the rest of her life, Frances would remember his dramatic antics at the pulpit.[50] She would also retain his dual emphases upon emotion and perfection, insisting that even the most depraved Americans could sweep aside their sins in a single, ecstatic moment.[51]

After the Civil War, the "Woman's Crusade" revived the drama and exuberance of antebellum revivalism. Starting in Ohio and spreading eastward like a "prairie fire," as Willard described it, the Crusade enlisted thousands of American women to sing, pray, and demonstrate outside of saloons. Historians have correctly credited Willard with moving the WCTU beyond this Gospel Temperance base and into politics, especially suffrage and partisan activity. Yet she also sought to imbue such efforts with "the spirit of the Crusade," as her longtime secretary Anna Gordon remarked. In her adopted state of Illinois, for example, Willard organized bands of women to sing "Home Sweet Home" at the legislature on behalf of a suffrage bill. After Willard allied the WCTU with the Prohibition Party, union members marched in prohibitionist parades, solicited "first voter" pledges for the party ticket, and decorated polls on election day. Through this new form of "Gospel-Politics," as Willard fittingly called it, the WCTU would fuse the evangelical methods of her girlhood to a fresh set of goals.[52]

To Mary Hunt, however, any such hullabaloo violated the Puritan canons of *her* youth. Echoing Beecher and other antebellum spokesmen, she condemned revival-style tactics in the quest for Scientific Temperance. To sway lawmakers, she counseled, use petitions and letters rather than parades and prayer-meetings; school officials, meanwhile, would be more easily reached by discreet visits instead of loud demonstrations. In part, these warnings represented a simple strategic calculation: like political partisanship, Hunt believed, marches and sing-alongs would only "awaken opposition" to STI. Yet they also reflected her Beecherite commitment to "cold logic" over "cold-water stories"—that is, to reason rather than passion. Central to this vision was Hunt's emphasis upon science, which would "leave no room for an emotional reaction." Others might set audiences ablaze with their "philippic against 'rum,'" as one STI ally noted, "but we are telling the children what 'rum' is, *scientifically considered.*"[53]

Willard and Hunt III: The Challenge of LTL

Frances Willard wanted to tell the children much more than that, as Hunt would discover. Like her Gospel-Politics, Willard's views on education fused the revivalist methods of her past to the new challenges of the present. Chairing the WCTU's first Committee on Juvenile Work, Willard suggested as early as 1876 that public schools teach "the effect of alcohol on the system." Yet she also called on WCTUs to forge their own "Juvenile Unions" against alcohol, reminiscent of antebellum youth organizations. "Reformed men should be invited to relate their experience," Willard wrote, praising "the noble Washingtonian movement" of the 1840s. "Temperance story-books should be read aloud. . . . Some simple badges or regalia will add much to the attractiveness of the Society." Four years later, as the WCTU's new president, Willard established separate departments of STI, Juvenile Unions, and Sunday schools. Scientific Temperance would attend to children's "intellectual knowledge" in public classrooms, as one Willard ally explained, while the WCTU's youth and religious schools would care for their "conscience."[54]

For a brief period, Willard gave equal attention to each wing of this "three-armed service," as another WCTU official described it. After she split with Hunt over the partisan issue, however, Willard started to assign "weight of influence" to Juvenile Unions. The first sign was Willard's appointment of Anna Gordon, her closest friend and confidante, as assistant superintendent of the Juvenile Unions department. The next was an 1884 national resolution, hailing "the legal enforcement of the scientific temperance law" but warning that "effort will not be relaxed in the formation of juvenile societies." Under its prior director, Julia Colman, the Juvenile Unions department had eschewed "enthusiastic" activities like parades and pledge signings. Now Willard and Gordon engineered their revival, drawing explicitly upon antebellum cold-water idioms. "To join our army you sign the muster-roll called 'The Total Abstinence Iron-Clad,'" Willard wrote. "Come, we want you to bear aloft a banner in your firm, young hands, and to inscribe upon it: '*Tremble, King Alcohol! we shall grow up.*'" Badges, songbooks, pledges, and military drills would make the entire effort "more of a social affair," Gordon added, attracting many more "young soldiers" than before.[55]

To underscore this new martial spirit, the WCTU reorganized its juvenile societies as Loyal Temperance Legions in 1886. Publishing its own weekly journal, the *Young Crusader,* the LTL department also distributed a "manual of arms" for boys as well as "broom brigade" drill instructions for girls. Gordon authored her own set of *Marching Songs for Young Crusaders,* which sold over 130,000 copies in just four years. Eventually, complaints from the WCTU's Peace and Arbitration department forced the LTL to abandon its military drills. Yet the legions continued to enlist children of both sexes, spurred on by Frances Willard.[56] Even as Willard and state WCTU presidents degraded her beloved STI, then, Hunt would have to ensure that local unions defended it.

Hunt's Response: Cultivating Local Allies

But how could she reach them? Edited first by Willard's sister-in-law and then by Mary Allen West, a close Willard associate, the *Union Signal* could not be expected to print Hunt's side of the story. From her Hyde Park headquarters, then, Hunt distributed broadsides, pamphlets, and magazines to WCTU women in the field. She also exchanged personal letters with thousands of these women, thereby short-circuiting her critics' favorite complaint. Since STI's inception, state WCTU officials had accused Hunt of violating "local" authority and prerogatives. Now Hunt turned the same argument against them, claiming that local STI superintendents—not state WCTU bosses—reflected the true spirit of Scientific Temperance.[57]

In Ohio, for example, Hunt mobilized district and county women against their own state STI superintendent, Frances Leiter. Hunt and Leiter split during Ohio's 1888 campaign for compulsory STI, which came on the heels of several statehouse setbacks. Fearing yet another defeat, Leiter dropped text requirements from her bill; indeed, the STI measure she submitted specified that "oral instruction" would satisfy it. The law sailed through Ohio's legislature but infuriated Mary Hunt, who had warned Leiter that schools could easily evade such a diluted measure. Hunt was correct. Districts proceeded to neglect or ignore STI, which "made many [WCTU] women question the utility of the law and made Frannie Leiter quite defensive about it," as one historian has written. In 1889, Leiter established an independent Ohio WCTU department with its own name—"Temperance Education"—and its own journal, the *Scientific Temperance Bulletin*. Halting the distribution of Hunt's *School Physiology Journal* to district superintendents, Leiter simply sent them her *Bulletin* instead.[58]

As in other states, however, loyal WCTU women corresponded directly—and secretly—with Mary Hunt. "Let me say to you (just in a whisper) I could not have effected much without the assistance of the National Superintendent," district STI superintendent Sarah Pratt wrote Hunt in 1891. Pratt enclosed twenty-five cents for one hundred copies of Hunt's "Schedule and Standard" on proper textbooks, noting that Ohio's texts would come up for readoption the following year. Yet few WCTU women were even aware of this deadline, Pratt said. "There is a missing link somewhere," she wrote. "The State should be 'snowed under' with the Schedule and Standard." At a minimum, Pratt promised, she would ensure that her own county superintendents received it. "Of course, this will give offence [*sic*] if it should come to the ears or eyes of the State Superintendent," Pratt told Hunt. "What course should District Superintendents take in this matter? You are at the head of this Department."[59]

The following spring, Hunt advised Pratt to send a series of her national STI circulars to county superintendents across Ohio. Pratt was reluctant to comply, fearing that other district WCTU officials would resent this obvious intrusion upon their turf. When she finally sent out the circulars, however, they sparked a massive grass-roots groundswell for Mary Hunt. "Though I have been County Superintendent of this Department for years I am ashamed that I had not thought of applying to [Hunt] before," wrote one woman. "Ohio has been defrauded. The expensive Bulletin issued by Mrs. Leiter is the only thing except a circular letter or two that has been furnished to superintendents. . . . [T]he little package sent me by Mrs. Pratt is of ten fold more value." The following year, a WCTU survey found that fewer than one-sixth of Ohio's schools required textbooks or regular exams for STI. Humiliated again, Frances Leiter conceded that she had erred by dropping the text provision from her earlier bill. But she still seethed inside, admitting that she "longed for revenge" against Mary Hunt. So did other foes like New York's Mary Burt and Illinois's Mary Allen West, both of whom tried—unsuccessfully—to interest local unions in Leiter's *Bulletin*.[60]

The View from the Field: Local Unions, STI, and LTL

They had better luck persuading unions to take Anna Gordon's *Young Crusader*, the weekly magazine of the Loyal Temperance Legion. "It is almost impossible to resist the importunity of a National Superintendent whose success depends in engrafting her specialty on the policy of each State," wrote Michigan WCTU president and Willard ally Mary Lathrap, in a thinly veiled 1891 attack upon Mary Hunt. She added a special plea for Loyal Temperance Legions, which an increasing number of local unions were selecting over Scientific Temperance. Of fourteen unions in Michigan's eleventh district, for example, eight had formed LTL departments by 1893, while only four bore departments of STI. The LTL emphasized "phases of the subject that cannot so well be taught in public schools," one enthusiast noted, especially "Biblical temperance" and the "moral" dimensions of drink. Most importantly, it represented "the best method of interesting the children in the temperance cause," as a rural New York WCTU resolved. Wary of yet another academic subject, young people preferred informal parades, poems, and pledge signings to dry scientific recitations at school.[61]

In her letters and literature, Mary Hunt warned unions against replacing their STI departments with the LTL.[62] Just four years after its formation, nevertheless, the LTL could boast almost 200,000 young members nationwide.[63] Some local WCTUs oscillated between the departments, promoting STI one year and LTL the next. Others combined them, sponsoring physiology classes for their Legions as well as parades and sing-alongs in the schools.[64] Indeed, unions found, an active Legion often solidified local sentiment behind STI. In Illinois's largely German Effingham County, one woman reported, school patrons were "intensely opposed to temperance teaching"; after the WCTU created a Legion, however, most of their antipathy melted. In northeastern Vermont, meanwhile, a union official noted that several of her town's schoolteachers were "graduates from the L.T.L."—and, by corollary, strong supporters of STI.[65]

Atop the WCTU, however, the two departments remained separate, hostile, and unequal. At Frances Willard's final WCTU national convention before her death, for example, a special "STI Evening" featured Mary Hunt leading local schoolchildren in recitations about "Wine and Cider," "Beer and Health," and "Alcohol versus Brains and Nerves." Elsewhere on the program, an entirely distinct "Young People's Night" entertained delegates with marches, pledge signings, and Willard's "favorite LTL songs." The same convention nearly halved Hunt's annual appropriation, confirming the antipathy of Willard and her clique. Behind the scenes, meanwhile, they were already conspiring with a fresh generation of experts—physiologists, psychologists, and educators—against Scientific Temperance. Whereas earlier challengers censured STI for violating local authority, these new foes would condemn it for undermining *their* authority. In the balance lay not just the destiny of STI, but the right of lay citizens to determine what—and how—their children would learn in school.[66]

PART II
RIVALRIES: FIGHTING EXPERTISE, 1891–1906

[W]e must fight fire with fire, physiological research with physiological research.
LEWIS D. MASON, 1903

[A]ll through the Bible, the blessings of right-doing are coupled with vivid descriptions of the consequences of following the wrong.
MARY H. HUNT, 1900

There is a certain set of men in educational circles [who] seem to think they own the schools, that the people, the parents have no right or voice in what should be taught their children.
MARY H. HUNT, 1899

Starting in the early 1890s, Mary Hunt and Scientific Temperance Instruction clashed with a new generation of experts. Trained in the laboratory, physiologists claimed that STI exaggerated the "poisonous" nature of alcohol. Citing their own advanced degrees, school officials charged that STI's scare tactics violated both their administrative discretion and the latest discoveries of academic psychology. Classroom teachers complained that the subject overloaded their schedules, bored their students, and insulted their own professional prerogatives. Following in the footsteps of Frances Willard, finally, a fresh set of leaders within Hunt's own Woman's Christian Temperance Union urged deference to these new expert foes.

Beneath the expert critique lay a distinctly liberal philosophy, dating back to the early nineteenth century but recast in a modern, technical idiom. Like antebellum evangelists, STI's enemies stressed emotion and passion over intellect and reason. Yet they cloaked these claims in the language of science, insisting that STI contradicted timeless principles of child development. They also abandoned their forebears' emphasis upon personal sin and salvation, emphasizing the public sphere instead. To these liberal critics, STI epitomized the inward-looking, anxiety-ridden "Puritanism" of their own youths. Directed squarely at each individual conscience, Scientific Temperance aimed to create a better society by converting each of its members. Just as STI neglected new scientific knowledge about alcohol, liberals said, so did it ignore new insights about the social roots of behavior.

In response, Mary Hunt sought to sustain Scientific Temperance through the same vehicle as her enemies: modern expertise. Fighting "fire with fire," as one supporter suggested, she enlisted several leading physiologists to vouch for STI's validity. Yet she failed to win similarly prominent allies from the world of education, drawing instead upon older languages of religion and democracy. Quoting Scripture, Hunt argued that God himself used fear to influence behavior; streaming back into American schools, meanwhile, her assistants reminded critics that "the people" favored STI as well. Even as she proclaimed STI's popular pedigree, however, Hunt also tried to establish a graduate college that would train expert instructors in the subject—and place it, once and for all, beyond debate. Liberal or conservative, every side to the controversy continued to presume that expert authority would settle it.

3

"WHEN THE DOCTORS DISAGREE": SCIENTIFIC TEMPERANCE AND SCIENTIFIC AUTHORITY

Like many of Mary Hunt's correspondents, the Reverend Merrill C. Ward of West Higham, Massachusetts, solicited her advice about "Science"—the science of alcohol. Reluctantly, Ward had agreed to lead a 1901 discussion on the subject before the Professional Club of nearby Brattleboro, Vermont. The discussion would follow a paper by C. S. Pratt, a local physician—and the leading focus of Ward's worries. "I fear," Ward wrote,

> Dr. Pratt will take the positions of Professor Atwater. . . . I wish how-
> ever to be thoroughly prepared to defend total abstinence, Prohibition,
> and the Text Books on scientific grounds. . . . I wish to know if there
> are *authorities* who will be recognized and *acknowledged* as such by *all
> parties* who are students of this matter. . . . Of course, I know there are
> scientific investigators of more or less reliability who have pronounced
> both favorably and unfavorably toward total abstinence, but is there a
> list that people of all views must admit as *higher authorities*? and con-
> sidering their opinions fairly and accurately where does the preponder-
> ance of evidence clearly lie?[1]

Ward sent a nearly identical letter several days later to Wesleyan University chemist Wilbur O. Atwater, again underlining the search for "*the authorities* on the effects of alcohol." This time, however, the minister omitted mention of his "position" on Scientific Temperance Instruction. "I wish to be truly scientific in temper," Ward emphasized, "and not express an *opinion* . . . but present briefly, the *facts* discovered by the most competent investigators."[2]

Together, these inquiries demonstrate Mary Hunt's key strategy in her strug-gle against scientist foes like Atwater: the discovery, recruitment, and promo-tion of scientist allies. Historians typically trace such schisms to the post–World

War Two polity rather than to the Progressive era, when disciplines divided
investigatory areas and demanded internal accord.[3] In the social sciences, pro-
fessorial entrepreneurs enforced this consensus by retreating from controversial
"reform" questions.[4] Physicians, meanwhile, coalesced around "science" itself—
around a new laboratory-centered research ethic that inflated their influence
well before it transformed their therapies.[5] Thanks to Hunt, however, equally
"scientific" and "competent" experts—in Merrill Ward's parlance—would assume opposite positions on Scientific Temperance. For every physician like
C. S. Pratt who skewered STI "on scientific grounds," Mary Hunt managed
to enlist a doctor—and, eventually, entire associations—who would support it
on these very grounds. She even secured a few allies within the American
Physiological Society, deacon of the new research gospel in American medicine.

Hence the STI controversy has been wrongly cast as a struggle for the
"public meaning" of science, pitting a nascent laboratory-based knowledge
against a "Christian physiology" of earlier vintage.[6] As their determined defense of vivisection will illustrate, Hunt's medical allies proudly proclaimed—
indeed, in some cases, pioneered—the new spirit of experimental study. Yet
in a culture that celebrated the moral efficacy of "science" before its practical
efficacy was established, scientists who created knowledge could trump doctors who consumed it.[7] Hence Hunt looked across the Atlantic, where several internationally known investigators affirmed the accuracy of STI and
offered additional evidence on the psychiatric—rather than strictly physiological—dangers of alcohol. Closer to home, Hunt also invoked the handful
of American neurologists and asylum superintendents who studied alcohol
addiction. In both cases, she hoped, new types of evidence would lend new
credence to Scientific Temperance.

Across America, of course, final curricular decisions still lay primarily with
thousands of local school districts. So Hunt sent reams of scientific literature
to her WCTU assistants in the field, encouraging them to publicize the
names and claims of "her" experts. She also prepared point-by-point refuta-
tions of prominent enemies like Wilbur Atwater, who spearheaded scientists'
campaign against Scientific Temperance. To be sure, Hunt's lieutenants were
occasionally loath to take up this new battle. Despite her best efforts to
simplify complex scientific issues, some WCTU women simply could not
comprehend them. Others lacked the confidence even to try. Phrases like
"higher authorities" and "competent investigators"—to borrow again from
Merrill Ward—seemed to imply that that laypeople were *incompetent,* at least
in "technical" matters like STI. Indeed, Hunt herself often skewered her
scientist enemies for pandering to an ignorant public.

Nonetheless, local WCTU women amassed enough information—and, in the process, enough assuredness—to sustain Scientific Temperance against its scientific critics. Pressing school boards to adopt Hunt's endorsed textbooks, they tried to show how these books reflected "the latest discoveries of science on the alcohol question," as one Ohio woman wrote.[8] Such activity sheds light on the vast appeal of "science" to the lay public, a key conundrum in Progressive-era historiography. We know far more about why doctors united around the new research ethos than we do about why laypersons increasingly accepted it.[9] Hunt and her assistants seized upon the ethos not because they were dazzled, confused, or mystified by it, but rather for the same reason that scientists and doctors did: to advance their own agenda.

In no case, we must emphasize, did this agenda include a broad public debate. Indeed, both experts and laypeople embraced "Science the Arbiter"—in Hunt's redolent words—because they believed that it would place STI *beyond* debate.[10] They were wrong. From the highest citadels of laboratory research to remote rural school boards and classrooms, Mary Hunt and her assistants sparked precisely the sort of deliberation that "science" was supposed to squelch. Although Hunt continued to insist that the laboratory dispensed a singular truth, then, the history of STI told a very different story: that scientific "truth" was fluid, plural, and always contested. Moreover, *laypeople* could contest it. Historians have correctly emphasized that laboratory research made science "something different," removing it from "common sense," the "local community," and "the ordinary experience of Americans."[11] Yet ordinary Americans—not just scientists—could exploit those differences, cloaking their older, commonsense experience in the new, allegedly invincible authority of the laboratory. In local communities, especially, Hunt and her WCTU assistants made "elite science" their weapon—and declared, against all evidence, that it would end all wars.[12]

Wilbur O. Atwater and the Committee of Fifty

America's medical and scientific communities paid scant attention to Scientific Temperance during its first decade. Several state boards of health denounced Hunt's endorsed textbooks for exaggerating the perils of alcohol.[13] Other physician groups rallied to STI's defense, noting the "poisonous influences" of liquor and "the dangers of even a moderate use."[14] An 1886 column in *Science* praised STI as a worthy "experiment," urging readers to suspend judgment until all results were in. By contrast, *Popular Science Monthly* editor E. L. Youmans condemned the new curriculum for its one-sided approach.

Alcohol represented "the most unsettled of all questions," wrote Youmans, a former champion of "scientific" teetotalism. "School-teachers can not explain it—doctors can not explain it; no two will agree about it."[15]

Nine years later, the Committee of Fifty to Investigate the Liquor Problem would convene to establish "a consensus of competent opinion" on the subject.[16] Boasting such luminaries as Charles Eliot, Seth Low, Daniel Coit Gilman, and Carroll D. Wright, the committee split into research squads that would study the legislative, economic, ethical, and physiological aspects of "the liquor problem." Chaired by John S. Billings, the Physiological Sub-Committee came to include Atwater, Harvard physiologist Henry P. Bowditch, Yale physiological chemist Russell H. Chittenden, and Johns Hopkins pathologist William H. Welch. All were members of the elite, inbred American Physiological Society (APS), the leading promoter of laboratory research in American medicine.[17] Like most APS members, each of these men hailed from the Northeast; each had imbibed the laboratory religion during graduate study in Germany; and—herein lay the true hallmark of the APS—each had assumed a full-time academic job upon his return, eschewing the private practices that were de rigueur for nineteenth-century medical professors. Together, they would mount the first sustained scientific assault upon Scientific Temperance.[18]

Atwater's laboratory "calorimeter" soon became the key arrow in the subcommittee's arsenal. Seven feet long, four feet wide, and six feet four inches high, this thermally insulated iron box was uniquely suited to capture the public's imagination. For Atwater would place not animals but human subjects in the contraption, measuring their intake of nutrients against chemical analyses of their trapped sweat, exhalations, urine, and feces. Performed with and without alcohol, such experiments promised "a more definite answer" to the question of whether alcohol was a poison (as Mary Hunt claimed) or a food. If the body released the same amount of heat in each instance and alcohol was not excreted "unchanged," physiologists could conclude that it was oxidized—and was, in this sense, a food.[19] Then, Atwater assumed, Hunt and her "extravagant agitators" would be dealt a crushing or even fatal blow. "In the teaching of temperance physiology," he wrote Billings in 1897, on the eve of his experiments,

> the points most at variance with physiological science . . . connect with the action of alcohol as food and as poison. If we can get answers as clear, decisive and authoritative as possible to the two questions, is alcohol food, and is alcohol poison, we shall do a great deal to clean up the situation.[20]

Yet even Atwater's stunning technological wizardry could not clean up this troublesome situation. He did find that alcohol was oxidized like other nutrients, neither a new nor a controversial proposition. The controversy emerged from his further claim that this property gave alcohol a "food-value," which seemed to obviate STI's "poison" admonition. As Hunt quickly realized, the scientists' argument hinged not upon mysteries but upon meanings—of words. The scientists defined "food" broadly enough to include liquors; Hunt's task would be to define it narrowly enough to exclude them.

Hunt's Response

Reflexively, Hunt's initial retorts stressed the majoritarian theme that had sustained STI in its initial clashes with hostile school boards. "The legislatures of thirty-five States, and the National Congress, have by law commanded that this study shall be taught in all their schools," she wrote in 1892, scorning the first wave of "scientific" salvoes. "Are they all wrong, and [the critics] right?" Yet she also worried that the public would "naturally" line up behind the "great names" on the Committee of Fifty—unless, as one ally noted, she could assemble an authoritative army of her own.[21] On the eve of Atwater's 1897 experiments, then, Hunt circulated statements by nine "eminent medical men" vouching for the accuracy of STI's endorsed textbooks. Atop the list stood Nathan S. Davis, a founder of the American Medical Association and dean of Northwestern University Medical School. Davis's remarks set the tone for the testimonies, stressing that "the most varied and accurate scientific experiments" exposed alcohol as "a subtle and deceptive poison." Each succeeding statement reinforced the research-based bedrock of total-abstinence teachings.[22]

As Hunt freely admitted, however, this circular only "silenced the clamor for a time." Atwater began publishing his studies in 1899, spawning a new STI search for authoritative allies. Hunt would continue to secure prominent physicians, tirelessly trumpeting their titles and attainments. But most of these medical men had distinguished themselves in the clinic and classroom, not in the laboratory. In the new prize ring of "scientific medicine," they could not pack the punch of a professional researcher. As Atwater and his calorimeter swaggered onto the American stage, then, Hunt set out to do what had seemed impossible two years earlier: drive a wedge through American physiology. "We must get the physiologists fighting among themselves, then the truth will come out," an adviser wrote Hunt. "Paul did this when he got the pharisees and Scribes fighting among themselves on the moral questions of the resurrection."[23]

In turn-of-the-century "elite science," however, Hunt encountered an enemy far more united—and far more hostile—than the litigious sophists of Scripture. Privately, researchers reserved their harshest invective for Mary Hunt and STI. "I wish you could get some of those cranks inside of your machine and cut off the oxygen supply," Columbia chemist C. F. Chandler wrote Atwater. "I am only afraid it is not big enough to do justice to the class of people who ought to be thrust into it."[24] In public, meanwhile, they confined themselves to turgid tributes to "science" and "fact." Physiologists, especially, seemed well positioned to achieve accord on both.

Allies I: Physiologists

Remarkably, however, Hunt managed to snatch several celebrities from the fold. The first was Harvard instructor Henry F. Hewes, a scientific stalwart who defended STI—and denounced Atwater—on the pages of the *New York Times* and other journals. Hewes argued that "food" should be defined not by oxidation but by "sum total effect"; thus narrowed, the category could not encompass alcohol.[25] Yet Hewes had earned his stripes in the clinic, not in the laboratory. In APS members Christian A. Herter and Winfield S. Hall, Hunt secured scientists par excellence on behalf of Scientific Temperance. A prominent Bellevue biochemist, Herter echoed Hewes's rhetoric regarding "food" but added a more technical retort to Atwater's secondary claim that alcohol "protects" body proteins.[26] Joining him was Northwestern's Hall, author of one of the earliest laboratory guides in physiology. Hall also coauthored an endorsed STI text series with Hewes.[27] All three defections had a distinctly Oedipal flavor. Hewes and Herter were protégés of Bowditch and Welch, respectively; Hall had studied in Leipzig with Carl Ludwig, mentor to nearly half of the founders of the APS.

Privately, colleagues groused that these "brother physiologists" had violated canons of "general courtesy." Committee of Fifty contributor and APS member Clifton Hodge admitted that he, too, had spotted some weaknesses in Atwater's research but had resisted revealing them, to avoid "anything which can be interpreted to show any bad blood or even annoyance on our side of the discussion."[28] Other APS stalwarts began to distance themselves from the Wesleyan chemist, suspecting that his forays into physiology lacked theoretical rigor. "Being a chemist," Welch wrote of Atwater, "he perhaps did not fully appreciate all of the physiological questions involved."[29]

To forestall further embarrassment, Atwater quietly withdrew his claim regarding protein protection.[30] At the same time, he fired back at Hunt's allies. The octogenarian Davis "is a most excellent man," Atwater wrote, "but he is one of the physicians whom the scientific men do not quite agree with." Specifically, Atwater recalled a survey asking medical school professors to pronounce upon "the so called germ theory of disease"; of all the respondents, only Davis "denied that disease was caused by bacteria." Since Atwater had not heard of Hunt's other medical spokesmen, he admitted, he could not comment specifically on them. But he felt confident that they, like Davis, were not "in accordance with the best of modern science."[31]

Like many professors and practitioners, Davis indeed doubted early bacteriological discoveries. Yet he quickly changed course and vouched for their validity, impressed as much by their political as by their curative potential. For in the germ theory and other laboratory-minted principles, Davis recognized the wherewithal for his life's work: the making of a modern profession, respected rather than ridiculed, with sanctified knowledge and certified knowers.[32] Contrary to Atwater's critique, then, modern science could encompass both the friends and the foes of Scientific Temperance. They coalesced under the banner of investigation, even as they bickered over its interpretation.

STI and Antivivisection

Here STI invites comparison with America's antivivisection movement, which appeared to challenge not just specific research claims but also research itself. In response, doctors rallied around animal experimentation with an unctuous unanimity that they never achieved on STI. True, "the commitment to medical progress through medical research was not shared by all physicians in this period," as historian Susan Lederer has written.[33] Handfuls of doctors did condemn vivisection, questioning not just the medical inferences but the moral integrity of the practice. Yet their voices were faint amidst the chorus of hosannas, emanating not just from "ultra-scientists" (as the antivivisectionists charged) but from rank-and-file physicians as well. Indeed, these general practitioners probably understood better than anybody else just what was at stake. The research ideal had rescued them from the ignominy of nineteenth-century medicine, especially from the degrading warfare between "regulars" and "irregulars." Antivivisectionism seemed poised to tarnish the lure of the laboratory—and with it, the potency of professionalism.

Yet Scientific Temperance worshipped this very laboratory—as the womb of teetotal teaching. Hence we should not be surprised to find Hunt, her medical enemies, *and* her medical allies supporting animal experimentation, perhaps the best illustration of their common ground within the contours of "scientific medicine." Turning a deaf ear to WCTU antivivisectionists, Hunt never hesitated to invoke animal studies if they seemed to support total abstinence.[34] Bowditch and Welch, meanwhile, played central roles in combating antivivisectionists.[35] But so did Davis. "If physiology has made any progress in any age," he declared in 1892, "such progress has been accomplished through experiment, and largely through experiments on animals." Davis even went on to praise French physiologist Claude Bernard, the bête noire of nineteenth-century antivivisectionism.[36] Winfield S. Hall, of course, did not need to assume an explicit position on the question; it could be gleaned from his laboratory booklets, which described operations upon rabbits and other animals.[37] Finally, medical societies that were ambivalent about defending Scientific Temperance were adamant about defending scientific experimentation.[38] Their position proved crucial in both cases, helping to explain why state legislatures so regularly required STI but so rarely restricted vivisection.[39]

Critics, of course, often linked the two movements.[40] At first glance, it was hard to deny the similarities. Both STI and antivivisection criticized scientists; both were dominated by women, especially by WCTU members. Yet antivivisectionists grounded their analyses in a larger critique of scientific *method*, which allegedly eroded religious and emotional impulses.[41] As evidence they cited the "savage materialism" of the "Continental laboratory," an unveiled jab at the alleged atheism and congenital cold-bloodedness of German-inspired research. A devout woman, Hunt would never acknowledge the first allegation. To the second, however, she was pleased to plead guilty as charged. Like her foes, she peppered her rhetoric with paeans to "the cold facts of science"—and especially to its handmaiden, the "Continental laboratory."[42]

Allies II: Psychiatrists

To trump American physiologists, meanwhile, Hunt looked across the Atlantic for a new set of expert allies: psychiatrists. "We who are leading the world in so many other things are sadly behind in this," Hunt told a WCTU gathering in 1901. "[W]hat European science has revealed . . . is beyond the knowledge of our university men who, nevertheless, think they know all there is to be known on the subject." Led by German teetotaler Emil Kraepelin, European psychiatrists had documented the damaging effects of alcohol on mental and muscular activity. Harnessing glossy new laboratory devices like the dynamometer and ergograph, these experiments promised to refute or at least amend Atwater's "food-value" pronouncements. His beguiling calorimeter had shown only that alcohol is oxidized, yielding energy; it did not clarify whether "this energy can be transformed into work," as one researcher wrote. So if science certified that drink inhibited muscular or mental ability—and if "food" included only those substances that enhanced this ability—then Atwater's entire edifice, like "the walls of Jerico [*sic*]," would come crashing down. "[T]he popular belief that alcohol imparts energy is a gross error," declared Belgian psychiatrist Jules Morel, in a typical passage. "Alcohol gives strength to no one."[43]

Published by the *American Journal of Sociology*, Morel's remarks caused at least one correspondent to send a worried letter to Wilbur Atwater. "When the Drs. disagree as widely as yourself and Mr. Morel," he asked, "what are we untutored folk to believe?" Atwater brushed off the challenge, acknowledging that "small quantities of alcohol" probably diminished "mental work" but also enhanced "muscular work."[44] Other physiologists sensed a more immediate danger. A nervous Committee of Fifty dispatched Johns Hopkins pharmacologist John J. Abel to undertake a "critical review" of the relevant literature.[45] Meanwhile, academic sparkplug G. Stanley Hall—like Abel, an APS member—began to publish articles on alcohol in his sounding board, *The American Journal of Psychology*. In these and other forums, critics noted that Kraepelin's results did not necessarily support rigid abstinence: whereas the inhibiting effects of large alcohol doses were beyond dispute, small amounts seemed to enhance both muscular and mental operations.[46]

Echoing Hewes's answer to Atwater, Kraepelin and his allies responded by emphasizing the cumulative rather than immediate consequences of alcohol. The "stimulating effect" vanished almost as soon as it arose, giving way to the "paralyzing effect" of fatigue—which "ultimately" rendered alcohol a depressant, one researcher wrote. Indeed, the meaning of "effect" could be expanded or contracted to yield whatever conclusion one desired. Like the quarrel about Atwater's calorimeter, then, this question soon turned upon definitions—of "effect," "stimulant," and of course "food." Under the influence of drink, "our apperception and consciousness, as well as the coordination of the movements, are disturbed," wrote Swiss psychiatrist and temperance activist August Forel, whom Hunt hosted during his 1899 visit to America. "I can believe as little in [alcohol's] pretended stimulating qualities as in its pretended nutrient power."[47]

Yet the very act of appealing to authorities across the Atlantic accentuated STI's weakness at home, as Hunt and her allies sensed. "The question of 'alcohol' will not be fought out in the German school of medicine but in the laboratory of the physiologist—or experimental physiology," one STI supporter wrote in 1903. "This is Atwater's *strong point* and we *must fight* fire with fire, physiological research with physiological research." So Hunt began a new search for an American expert, someone "who has studied the Alcoholic Question from the standpoint of recent investigations made on the subject in Europe"—that is, from a psychiatric standpoint. To Hunt's chagrin, however, no such scientist stepped forward. Stanley Hall had also hosted Forel during his American visit, finding him a "splendid man but a little extreme"; likewise, William Welch dismissed Forel as "the leader of the extreme 'Temperenzlers'" in Europe. In their view, the teetotal advocacy of foreign psychiatrists compromised their scientific objectivity—at least on questions surrounding alcohol.[48]

Allies III: Inebriests

Hence Mary Hunt turned to perhaps the only indigenous Gilded Age American science, the science of inebriety, which started in the United States and spread eastward to Europe rather than vice versa.[49] On top of their national origins, inebriests offered STI another obvious weapon: a scientific statement of alcohol addiction. Even enemies like Atwater had to admit that alcohol "has the power to create a diseased appetite for more," as he observed in 1898. Yet they could not claim to understand how, when, or why this occurred. Thus they often concluded their discussions with a vague caveat about inebriety, underscoring the confusion that surrounded it. "It may, perhaps, be said with safety that in small quantities [alcohol] is beneficial, or at least not injurious," wrote APS notable William H. Howell, *"barring the danger of acquiring an alcoholic habit."* [italics added].[50] Inebriests held out the hope of dressing this danger in the habit of science.

The danger, as such, was not new. Roughly a century earlier, Benjamin Rush had described inebriety as a "disease" that doctors could both recognize and remedy. Rush, Samuel Woodward, and a small core of other activists pressed for "moral treatment" of alcoholics within separate asylums.[51] But most of these institutions did not arise until after the Civil War, when they tended to assume one of two guises. Inebriate "homes" were private and voluntaristic: they stressed the spiritual poverty of the drunkard, whose transformation hinged upon his own cooperation. Inebriate "asylums" were usually public and coercive: like facilities for the insane, they emphasized the biological and especially the hereditary roots of inebriety. By the 1880s, the somaticists had squashed the spiritualists and assumed authority over American alcoholism. Leading the charge was the American Association for the Cure of Inebriety (AACI), a small group of neurologists and asylum directors who trumpeted the "medical" rather than "moral" nature of the disease.[52]

Fresh from this fracas, the AACI initially eschewed Mary Hunt and Scientific Temperance. In an 1886 editorial, it acknowledged that STI laws represented "a movement in the right direction." But "real responsibility rests on the medical profession, not on moralists and clergymen," the AACI added. "It is a scientific subject, that requires a medical training to study and determine." Inebriests especially objected to texts authored by "non-experts" like Mary Hunt, who cast the alcohol problem "simply as one of education." Upon inspection, however, it turned out to be as much a function of inheritance as instruction. Inebriests seized upon hereditarianism for the same reasons as American psychiatrists did: in the absence of any other established etiology, heredity alone seemed to hold out hope of illuminating discrete causes for mental disease. Yet heredity also clashed with Hunt's rosy rhetoric, which held that STI could "save the children" before they drowned in drink. At first glance, STI seemed like a "grand work of education," inebriest Lewis D. Mason admitted in 1888. "But will these measures restrain or restore a will degenerated and weakened by the excesses of 'alcoholic ancestry'?" he asked. "Are there not those who become 'inebriates from inheritance,' from 'necessity' not from 'choice'?"[53]

Inebriests never completely discarded this rigid determinism. Yet they soon realized that an exclusive emphasis upon heredity underscored their impotence, reminding observers that even these specialists could not cure the consequences of "alcoholic ancestry." Like psychiatrists, then, inebriests eventually downplayed both heredity and cure in favor of prevention. Shades of this shift had appeared in the late 1880s, when the AACI changed its name to the American Association for the *Study* and Cure of Inebriety (AASCI). By 1895, the conversion was complete. "The cure of a few hundred inebriates in asylums will be lost in the larger questions of prevention," the AASCI declared. "This is the direction of scientific advance."[54]

So inebriests switched course and embraced Scientific Temperance, which likewise stemmed from the conviction that "success would be in prevention rather than cure."[55] Indeed, former critics like Mason and Hartford asylum superintendent T. D. Crothers would become STI's staunchest medical allies, serving on Mary Hunt's advisory board and staving off physiologists' salvoes. Meanwhile, inebriests' new preventive emphasis helped swell their own status, as reflected in their frequent contributions to journals and their flashy professional titles. Crothers, for example, abandoned his superinten-dency for a neurology chair—"professor of nervous and mental diseases"—at the New York School of Clinical Medicine. Lest anyone miss the point, the AASCI announced that his appointment signaled the ascension of inebriety "into the realm of scientific medicine."[56]

Rather than battling for the public meaning of science, then, STI's allies and enemies fought within a common definition—within, that is, a "scien-tific medicine" broad enough to accommodate any number of competing analyses by physiologists, psychiatrists, and inebriests. Yet each camp still clung to the dream of "a consensus of expert opinion," so "authoritative . . . as to make it practically irresistible," as Wilbur O. Atwater wrote in 1897. In 1902, MIT biologist and perennial STI critic William T. Sedgwick called for "a body of scientific men" who would resolve the alcohol question, once and for all; the following year, Mary H. Hunt proposed "a council of expert physiologists" that would do the same. After the Committee of Fifty released its anti-STI report, finally, Hunt's aides spoke of enlisting "as many as fifty authorities on our side."[57] Even in the heat of battle, then, each party re-tained the faith that laboratory research would settle it. Out in the schools, though, science was generating exactly the sort of freewheeling, broad-based discussion that both sides hoped it would stifle.

"Heated Discussions": Hunt, Science, and Local Debate

In March 1904, local STI worker Mary L. Brumbach accepted an invitation to address a public discussion on Scientific Temperance in Lexington, Illinois. The meeting was prompted by widespread news reports about the Committee of Fifty, which had recently released its study condemning STI as "neither scien-tific nor temperate." Now, Brumbach wrote Mary Hunt, "Teachers, Ministers, and others" were calling upon her to answer these charges. Brumbach had already received Hunt's Reply to the Physiological Sub-Committee, replete with explanations of the committee's "fallacies" and quotations from competing authorities. Pleading for more literature—"any and all helps . . . that will be of service to me"—Brumbach also asked Hunt for her prayers. "I am going in the fear of the Lord, and not man, and will do my very best," Brumbach declared. "I begin to feel that Illinois [will] be the battleground."[58]

A few weeks later, a triumphant Mary Brumbach wrote again to report on the meeting. Thanks to Hunt's aid, she had managed to persuade most of her listeners that STI still accorded with "the best science," Brumbach gushed. Although pockets of dissent remained, of course, the entire town's attention had been piqued. "The discussion called out a good sized audience of *intensely interested people,* largely women," Brumbach underlined, "and nearly every one present asked questions." Formerly skeptical of STI, local teachers seemed altogether more friendly. Best of all, the town's teacher-institute conductor—"violently bitter, unreasonable, and of the 'Atwater Stripe'"—had been "quietly retired." His successor could not attend the meeting, Brumbach added, but "he was reported to me as *not opposed* to STI."[59]

Not every STI worker was this successful in stalemating the scientific challenge, of course. Yet Brumbach's letters help illuminate the struggle's local dimension, so often obscured in our historical discussions of science during this era. Following the 1899 publication of Atwater's first calorimeter claims in an obscure Department of Agriculture bulletin, Hunt distributed a point-by-point sixteen-page rebuttal to over 45,000 teachers, ministers, journalists, and WCTU women. When the Committee of Fifty finally released its 1903 report, lending Atwater a new scientific imprimatur, Hunt organized an even broader grass-roots response. Composing an exhaustive twenty-six-page reply, Hunt directed state STI leaders to send it to legislators, school officials, and university professors; county STI leaders, to teacher institutes and medical societies; and local STI women, to lawyers, ministers, businessmen, and especially teachers. All told, nearly 200,000 copies of Hunt's *Reply to the Physiological Sub-Committee* circulated across America. "[T]he Report contains inaccuracies and misrepresentations that many people will accept simply because of their confidence in . . . the Committee of Fifty," warned Hunt, "unless we point out to them these inaccuracies and misrepresentations."[60]

Most of all, Hunt emphasized, WCTU women had to ensure that school textbooks still reflected STI's original total-abstinence principles. Hunt's endorsed texts had competed for years with so-called wet competitors, which omitted her claim that alcohol was poison. As Atwater also took issue with this premise, news of his research sparked a fresh round of battle in the "textbook war." Publishers flooded the market with wet texts, many of them invoking Atwater, Bowditch, and other Committee of Fifty physiologists. Across the country, then, WCTU women streamed back into school boards to block the new "Atwater books"—and to secure their own. "We must fight Overton until we drive it from the State, as it does not teach the total abstainance [*sic*] of alcohol," wrote Ohio's STI superintendent, referring to an especially popular wet volume by New York physician Frank Overton. "[W]hen our school men have had their attention drawn to this matter they will correct the mistake and put into the hands of our children newer and better books." To prepare women for such negotiations, Hunt disseminated pamphlets detailing Overton's defects. She also circulated hundreds of copies of the book, highlighting each objectionable passage.[61]

In the few states where central text commissions selected books, meanwhile, WCTU women bombarded these panels with petitions, letters, and samples of Hunt's "endorsed" choices.[62] Yet they also continued to pressure local boards, which frequently retained final discretion in the matter.[63] Since states were forever altering their textbook rules and requirements, moreover, many school districts were slow to adjust. Five years after Connecticut state school officials revoked their approval of a famously wet book, for example, STI worker Eliza Bliven found that classrooms in her native Canterbury still used it. Presenting several Hunt-approved texts to a member of her local school board, she asked to review a copy of the notorious "State Physiology." The schoolman hesitated at first, but finally he let her take one home. "I examined it last evening, and was shocked at its worthlessness," Bliven wrote Mary Hunt. "The part about liquors recommends them as a *food* [and] does not discountenance moderate drinking. . . . I wonder in how many towns of Connecticut, the children are being misled by this book?"[64]

Elsewhere, too, WCTU women reported a steady stream of wet books in their schools. In the decade following Wilbur Atwater's first experiments, American publishers released over twice as many wet books—that is, books labeling alcohol a "food"—as dry ones. Thanks to the efforts of local soldiers like Eliza Bliven, though, the WCTU's endorsed texts maintained a rough parity with "Atwaterism (should I say 'Atwhiskeyism?')," as another jocular Connecticut ally told Mary Hunt. In Massachusetts, for example, an 1897 survey of 115 schools found that 55 selected Hunt's books and 60 did not. Meanwhile, 78 of 88 counties in Ohio reportedly used the much-despised "Overton" series. Yet the WCTU successfully fought to remove Overton from the largest single book market in the country, New York City.[65]

Such activity only heightened the worries of scientist critics like MIT's William Sedgwick, who was quick to raise the specter of censorship. Defending his friend Wilbur Atwater at a 1901 debate with Mary Hunt, Sedgwick charged that Hunt and her "self-constituted oligarchy" of WCTU assistants conspired to keep Atwater's research out of American textbooks. Whatever their motives, however, Hunt and her aides clearly failed to "censor" contrary viewpoints from books—or from schools.[66] Indeed, textbook adoptions were often the occasion for "heated discussions" among "teachers, pupils, and patrons," as one Minnesota school official reported. Much to the WCTU's chagrin, moreover, the discussions frequently made a community more—not less—skeptical of STI's teetotal claims. "Parents take an interest in the subject," wrote one worried STI worker from Amsterdam, New York, "and some of the boys and girls investigate whether the scientific part as taught in school is true." Although they never intended or even envisioned such debate, finally, WCTU women could use it to comfort themselves in the wake of defeat. "It has set the people to *thinking* and *talking*, at least," wrote another New Yorker, after noting dejectedly that her own schools still rebuffed her pleas for STI.[67]

"I Would Not Know What to Do": STI and Lay Ignorance

In some communities, though, the very defenders of Scientific Temperance proved unable to comprehend it. Hunt warned WCTUs against appointing "noticeably uneducated" STI superintendents, who would provoke "just criticism" from her foes. Yet too many local unions still chose marginally literate women for the post, as a Michigan critic told her 1891 state WCTU convention. Two delegates proposed that unions require all STI superintendents to spell "scientific" and "ballot" correctly, sparking an angry rejoinder from a third woman who noted that "some of our best workers cannot spell well, but [that] does not advise putting them out of office." As a compromise, another speaker suggested that superintendents should be able to recite the state's STI law by heart. But this too was beside the point, a fourth critic noted. The real problem was a widespread lack of knowledge, she argued, not of literacy: among WCTU women, "not one in one hundred . . . locate the organs of the body properly." Indeed, a Pennsylvania STI advocate admitted, most union members "know no more about physiology than a child."[68]

Fearing public exposure and ridicule, such women were often reluctant to take up the struggle for Scientific Temperance. Whereas some unions appointed unqualified women for the STI helm, then, many others left it vacant for lack of volunteers. "This department, I think, has been hindered by its rather high sounding title," Maryland's STI superintendent complained in 1888, noting a shortage of local assistants. "The very sound of the word scientific seems to imply something deeper than the surface, and more difficult of comprehension than the usual routine of woman's thoughts."[69] With the rise of the laboratory, of course, science moved even farther below the surface—and beyond lay understanding. Unlike earlier textbook disputes, which involved only general principles of anatomy and physiology, the new controversies hinged upon specific, highly technical experiments.[70] Instead of seeking to master the complex world of calorimeters and dynamometers, then, some local unions simply conceded defeat. "We ought to have a Superintendent of

S. T. I. but there is no one who will take it," one rural New York WCTU president told Mary Hunt in 1902. "I cannot do everything, and besides I would not know what to do any way."[71]

Dismissing such complaints, Hunt continued to insist that "any woman of intelligence" could master the science of Scientific Temperance.[72] Yet her own paeans to professional expertise suggested otherwise, as several critics noted. Indeed, Hunt herself often skewered STI's scientific opponents for pandering to an ignorant public. When Norwegian explorer Fridtjof Nansen returned from his Arctic expedition, a *Union Signal* editorial noted, "he did not hurry from one gathering to . . . another" like Wilbur Atwater, demanding the ouster of textbooks that neglected his newfound knowledge. Instead, Nansen "went before bodies of scientific men—of specialists in geography—revealed his data and demonstrated the correctness and tenability of his conclusions." This "method in the propagation of new discoveries," the *Signal* concluded, "is the same method which scientific men generally use." To act otherwise was nothing less than "a breach of professional courtesy," as Hunt argued elsewhere. "Have we not been carefully taught," she asked, "that only a medical man is competent to answer medical questions?"[73]

Yet this same argument rendered Hunt and her STI lieutenants *incompetent*, since they lacked medical training and credentials. Opponents took special glee in ridiculing STI textbooks that Hunt had helped rework along total-abstinence lines.[74] "It is quite evident that she was scarcely qualified to undertake the work of revising," a Pennsylvania doctor charged in 1902. "These books should be supervised by medical men and not by the laity." Atwater ally William Sedgwick went a step further, charging Hunt with subjecting "modern science" to lay "propaganda." In response, Hunt told Sedgwick that "the mothers in any community" had "a perfect right" to select their children's textbooks. But if "medical men" constituted the only proper court of appeal, "the mothers"—including, most obviously, Mary Hanchett Hunt—lacked *any* right over Scientific Temperance Instruction.[75]

"The People Are Quite Competent": STI and Lay Knowledge

For the most part, then, Hunt avoided purely scientific or majoritarian arguments in her battle against Wilbur Atwater and his fellow physiologists. Instead, she combined them by insisting that lay citizens could comprehend and apply expert prescriptions. "With all due respect to the culture of Middletown," she quipped, in reference to Atwater's Wesleyan experiments, "happily there are in the United States, outside of that city, some other people able to understand technicalities." To be sure, many laypeople "still cling to the belief that science consists chiefly in big words," as one Hunt ally noted. Yet if science was "stripped of its sesquipedalian garments," he added, anyone could grasp it. "The people are quite competent to weigh all the arguments pro and con," Hunt declared in 1901, "when put before them in plain terms that will not befog."[76]

Privately, Hunt sometimes wondered how much laypeople could actually comprehend. "The great public," she told physiologist ally Winfield S. Hall, "is too busy to stop to consider the merits of the case." This was especially true for WCTU women, whose domestic chores often precluded in-depth scientific study.[77] To supplement her lengthy, point-by-point responses to Atwater and the Committee of Fifty, then, Hunt also prepared a series of seventeen short pamphlets, between two and four pages in length, each answering a common anti-STI attack. Titles included "Does the Oxidation of Alcohol Prove it a Food?" "Is Alcohol a Force Regulator?" and "The Effect of Small Quantities of Alcohol on Mental Operations."[78]

Eschewing the dense scientific argument of Hunt's other writings, these pamphlets emphasized the names of her prominent scientific allies instead. "Perhaps you do not know just what to answer when the opponents of temperance education say, 'There is no scientific basis for total abstinence,' and that 'it is not right and just to teach that there is,'" Hunt wrote, introducing the new pamphlets. "You feel sure such critics are wrong, but you do not know exactly how to prove their error. . . . They *are* wrong. The greatest scientists in the world have proved all such criticism to be erroneous." Here Hunt could draw upon her famous card catalog of scientists' quotations, "ready for any emergency that needs to be met." Whenever Atwater, Sedgwick, or any other antagonist raised a new objection to STI, Hunt would dash off a new pamphlet—peppered with remarks from "supporting authorities." In 1901 alone, Hunt and her five full-time secretaries distributed 160,000 such leaflets to women in the field.[79]

At first glance, it would seem that these leaflets' emphasis upon expertise also discredited lay experience and authority. In practice, however, they motivated WCTU women to acquire even more expert knowledge—and to assert it as their own. For as Hunt often emphasized, her literature was designed not simply for "information" but for "preparation": that is, to ready local STI workers for face-to-face encounters with school trustees, principals, teachers, and students.[80] Well before Hunt circulated any responses to Atwater, for example, a Denver STI worker wrote for assistance in composing her own rebuttal to present in her schools. Women in Ann

Arbor, Michigan, scoured their local newspapers for alcohol-related statements by "prominent physicians," presenting a scrapbook on the Atwater
controversy to area teachers. As their state text commission met to select an
STI series, meanwhile, Tennessee WCTU women held meetings to study
dry books and their wet rivals. They also prepared a circular comparing
these options, which they distributed to teachers and other "prominent
people" throughout the state.[81]

In circular fashion, finally, such knowledge gave STI workers the strength
and confidence to continue the debate—even against far more powerful opponents. In Brooklyn, for example, prominent minister L. P. Armstrong was
astonished when a WCTU congregant stood up to challenge his own statements in support of Atwater. "I said that 'while it would be unwise to
mention those things to ignorant children, I had no doubt that your conclusions were as accurate as your investigation was sincere,'" Armstrong wrote
Atwater. "A lady, who by the way is a friend of Mrs. H.—took issue with
me. . . . [S]he read from a pamphlet, not yet in the hands of the Public,
which Mrs. H. is preparing." Elsewhere, too, Hunt's literature helped steel
WCTU women against the often shrill attacks of enemy experts. "I recommend that not one white ribboner in all Michigan set up a lamentation
because all men do not always, and everywhere speak well of us," declared a
Michigan WCTU official in 1904, reviewing recent struggles over Scientific
Temperance. "Let . . . the editor of The Outlook and 'The Committee of
Fifty' and all the rest of them have their little say."[82]

As these remarks suggested, laboratory physiologists were not STI's only
expert foes—or even, necessarily, its most dangerous ones. On the pages of
The Outlook, America's leading liberal church journal, editor Lyman Abbott
published numerous columns condemning Scientific Temperance. Echoing
Atwater and the Committee of Fifty, a small handful of these pieces charged
STI with perpetuating scientific falsehoods about alcohol. More commonly,
though, Abbott—like L. P. Armstrong—blasted STI for violating "vital principles of sound pedagogy." Allies, too, emphasized the shifting nature of
these attacks. "The battle for truth has lost nothing in this controversy thus
far, but the opposition is not dead," warned the Union Signal in 1901. "[I]t
is changing base to so-called 'pedagogical lines.'" In her own national report
for that year, Mary Hunt also warned STI workers to stay on their guard.
"THE DOCTORS FAILED THE BREWERS," the report screamed. "WILL
THE TEACHERS?" In the balance lay not the "physiological effects" of
alcohol, but rather the issue of how—or even whether—American children
should learn about them.[83]

4

"LET THE PEOPLE DECIDE": SCIENTIFIC TEMPERANCE AND EDUCATIONAL EXPERTISE

In 1896, the superintendent of America's largest school system went before his state legislature to denounce Scientific Temperance Instruction. Charles R. Skinner charged that New York's strengthened STI law "deprived school authorities of the fundamental right to regulate courses of study," because it added "minute specifications" for textbooks, examinations, and a minimum number of lessons. Even worse, Skinner maintained, STI reflected the ancient fallacy that knowledge would influence behavior. Lacking mature intellects, however, young students could not comprehend detailed anatomical discussions or deduce a course of action from them. Here Skinner cited the new psychological discipline of "child-study," a "thoroughly scientific development" that allegedly invalidated Scientific Temperance. For if young minds were essentially innocent, as child-study theorists insisted, frightful warnings could only harm them.[1]

A few weeks later, STI stalwart Albert H. Plumb took aim at this rising pedagogical challenge. A Boston minister and member of Mary Hunt's advisory board, Plumb noted that "multitudes of educational experts" still vouched for STI's knowledge-driven approach. In the same breath, however, Plumb invoked a still higher authority on behalf of Scientific Temperance. Throughout Scripture, Plumb maintained, "the Great Teacher himself" confirmed the crucial impact of information and understanding upon human action: "Sanctify them through Thy truth," "Thy word is truth," "My people are destroyed for lack of knowledge," and "The truth shall make you free." Finally, Plumb noted, STI bore a popular imprimatur that its critics never earned. "[T]he people of this country, the parents of our school children, have declared that [STI] shall be thus taken up," Plumb concluded. "And how utterly wrong-headed [for] the servants of the people, or for anyone else, to interfere." Professional, theological, and democratic

authority all stood squarely behind Scientific Temperance, recent educational attacks notwithstanding.[2]

But the attacks would continue, with greater force and frequency. By
1900, in fact, nearly every leading American educator opposed STI. In many
ways, their critique echoed the romantic cadences of antebellum evangeli-
cism: to child-study advocates, passion and intuition—not reason and intellect—paved the road to Heaven. But they clothed this philosophy in the new
garb of science, insisting that STI violated timeless principles of human
development. Like laboratory physiologists, they also claimed an exclusive
right to investigate and interpret their subject. Whereas an earlier generation
of schoolmen had welcomed Scientific Temperance as another step in the
struggle against local authority, then, this new generation of experts condemned it for trampling upon *their* authority—as "scientific" educators.

In response, Mary Hunt sought to combine three separate and sometimes
contradictory strategies, neatly summarized in Albert Plumb's 1896 remarks.
As in her battle with rival physiologists, Hunt initially tried to fight fire with
fire by enlisting her own experts.[3] Like its critics, however, STI's educational
allies displayed a deep antipathy toward lay influence in school affairs. Hence
even friendly schoolmen were hesitant to lend their names to Hunt's agenda,
forcing her to invoke an even older authority: the ministry. Here, too, Hunt
sought to enlist prominent clergymen who would vouch for STI's conservative insistence that knowledge of evil could cause individuals to cast it out.
Yet God's Word was also subject to a myriad of interpretations, as a rueful
Mary Hunt would discover. Imbued with complex techniques of "higher
criticism," a new generation of liberal theologians read the Bible as a set of
allegories rather than literal truths. To these ministers, STI reeked of the
same sterility and scare tactics as the Puritan orthodoxy they had left behind.
In religious as well as educational circles, it seemed, STI's allies lacked the
expert pedigree of their foes.

To defend STI, then, Mary Hunt would revive the same majoritarian
appeal that had powered her initial sweep through American legislatures. In
states where her STI law appeared in jeopardy, she organized conservative
churchmen into "central committees" that distributed literature, collected
petitions, and lobbied legislators. Lest STI lose ground in the classroom,
meanwhile, she also called on local WCTU workers to converge yet again
upon superintendents, principals, and teachers. In each case, supporters voiced
a simple theme: "Let the people decide."[4] Yet the plea often clashed with
Hunt's erstwhile tributes to "eminent experts," "competent professionals,"
and "the best men." The more that citizens deferred to educational authori-

ties, in fact, the less likely they were to support Scientific Temperance. Even within Mary Hunt's own Woman's Christian Temperance Union, she would discover, vast numbers of Americans were embracing liberal educators—and denouncing STI.

"Hearty Cooperation": STI and State Educators, 1879-1894

From its birth, most educational officials backed Scientific Temperance. Enshrined in state law, STI echoed the centralizing impulse that had marked their movement since the days of Horace Mann. It also mirrored Mann's emphasis upon the assimilative power of common schools, which would fuse disparate American youngsters into a single, moral whole. "[F]ormation, not reformation, is now the educational watchword," declared National Education Association president Thomas Bicknell in his 1884 address. "[STI] shall make us the saviors of little children." After a rousing speech by WCTU president Frances Willard, the NEA's "special temperance committee" recommended "the hearty cooperation" of American schoolmen "in making this legislation general throughout the land." Several years later, after most states passed STI laws, a second panel of eighteen state and city school superintendents endorsed Hunt's "standard of enforcement"—including a minimum number of lessons, regular examinations, and textbooks "in the hands" of pupils. Scientific Temperance "brings a strong moral sentiment to the support of the schools," one state official remarked hopefully, "establishing them more firmly . . . in the confidence of the public."[5]

Amid this chorus of praise, a few educators did criticize STI. Some worried that Scientific Temperance would actually erode "the confidence of the public," by alienating besotted parents and stingy school boards. Others argued that STI neglected students' emotional natures. Temperance "is not to be accomplished by mere text-books and catechism teaching," one schoolman wrote in 1887, "but by a wise culture of the heart." The most prominent critic was Massachusetts reformer and Civil War hero Thomas Wentworth Higginson, who feared that STI would provoke a "morbid curiosity" about alcohol and other vices. In matters of sex and drink alike, Higginson wrote, "innocent ignorance is usually a better safeguard than precocious knowledge."[6]

During STI's first decade, though, such opinions were still in the minority. Rebuking Higginson, an editorial in Bicknell's *Journal of Education* reaffirmed the profession's overwhelming support for Scientific Temperance. "Even the knowledge of the sex functions will be imparted to our children long before they leave the school, either by kind parents . . . or by *vicious companions*," the editorial warned. "Which would Colonel Higginson prefer? They cannot be kept in ignorance on these subjects. The only question is, How shall they be taught?" For most leading schoolmen, the answer was obvious: by regular lessons in physiology and hygiene, with "special reference" to alcohol. That STI "had its origin in the mother-heart"—more precisely, in the WCTU—only reinforced its worth, according to its allies. "The two words—WOMAN and TEMPERANCE—each the symbol of the true and good, shall be forever united," Bicknell proclaimed, in praise of STI. Spanning the nation, Mary Hunt and her female legions would bind Americans more firmly to each other—and, even more important, to God. Here they resembled no one so much as the educational leaders themselves, who crisscrossed the country like itinerant pastors to promote common schools in God's name.[7]

The Educational Trust

After 1890, however, a new generation of school leaders came to the fore. In dozens of states, legislatures replaced urban ward-based trustees with small, citywide school boards. The boards, in turn, vested broad authority in newly minted "expert" superintendents. Often boasting graduate degrees in psychology or education, these schoolmen emulated engineers rather than evangelists; their watchwords were science and efficiency, not Scripture and enlightenment. "I should as soon think of talking about the democratization of the treatment of appendicitis [as] the democratization of schools," declared Nicholas Murray Butler, the first president of Columbia University's Teachers College and perhaps the best-known spokesman for this new breed of educator. "The fundamental confusion is this: Democracy is a principle of government; the schools belong to the administration; and a democracy is as much entitled as a monarchy to have its business well done."[8]

Men like Butler retained the millennial optimism—the "moral earnestness and sense of mission"—of nineteenth-century educators. Yet they rejected the older vision of redemption via democratic participation. Municipal reformers outside the schools would struggle to reconcile this ideal with government by experts, the pivot of their new bureaucratic politics. They often seized upon interest groups, which would represent citizens indirectly via new functional units—and would coordinate with credentialed, professional administrators to construct a frictionless utopia. "These were the years," concludes Richard McCormick, "when governments . . . began to take explicit account of clashing interests and to assume the responsibility for adjusting them through regulation, administration, and planning."[9]

Schools were the great exception. Educational reformers displayed the same devotion to expertise, of course—but without the corresponding commitment to group participation. "The leaders of the intellectual life of the city will have to evolve a plan," wrote Andrew S. Draper, superintendent of several school systems, "and the masses will have to be educated to its support." Other urban reformers, like McCormick, might speak of "formally recognizing and adjusting group differences" among segments of the population; schoolmen would speak only of winning their compliance. Policy itself was often forged by a small network of academics, foundation researchers, and superintendents that placed friends and students in offices, shaped legislation, and otherwise governed American schools for the next half-century.[10] This "Educational Trust" did not lack critics, of course: teacher leaders like Margaret Haley blasted its hierarchical organizational reforms, while dissident theorists like John Dewey and George Counts denounced the mechanical, formalistic curricula favored by many of its members. Yet even radicals who opposed the trust's "undemocratic" style of educational policy seemed to support—or at least accept—its undemocratic style of educational *policy making*.[11]

Educators Versus "Enthusiasts"

This new generation of schoolmen censured STI as loudly as earlier educators had supported it. A good measure of STI's falling favor was its rude reception at the National Education Association, which had embraced it so eagerly just a few years before. In the interim, a fresh coterie of experts had replaced genteel Mugwumps like Bicknell on the NEA board of directors. Studded with luminaries such as Butler, Chicago superintendent Albert Lane, and Kansas City stalwart J. M. Greenwood, the board rejected Hunt's 1894 plea to create an STI division within the association. Thereafter, three NEA curricular committees—the tediously titled Committees of Ten, Fifteen, and Twelve—snubbed STI in rapid succession. Mary Hunt personally lobbied the panels, assembled to design courses of study for secondary, primary, and rural schools, respectively. Yet all three committees refused her request to include STI as a regular subject. By 1900, a much-heralded NEA "roundtable" on STI turned into a rout. Just two educators at the event supported Scientific Temperance, while all the other schoolmen—including legendary Chicago Normal School principal Francis Parker—attacked it.[12]

To Hunt's horror, even female school officials began to denounce STI. As recently as 1888, an Illinois convention of "school mistresses"—professors, superintendents, principals, and "leading teachers"—unanimously endorsed the subject. New York's WCTU dotted its annual report with statistics on schoolwomen, assuming that female officials would be more likely to enforce Scientific Temperance.[13] As the century drew to a close, however, this presumption melted away. Growing numbers of women assumed state and county superintendencies, especially in the West; but they skewered STI as fervently as their male counterparts.[14] Even at local levels, schoolwomen—like schoolmen—often opposed STI. "Principle [*sic*] is not a believer in Scientific Temperance Instruction Books," reported an outraged WCTU school visitor in Burlington, Vermont. "She does not use them, or require the Teachers to do so. . . . She believes in simply ignoring the subject as far as possible."[15]

For this new cohort of educators, STI's popular imprimatur was irrelevant. Scientific Temperance should be left to experts, one schoolman declared, since they "know best what subjects are suited to the different stages of mental and moral development" in children. Some educators conceded that legislatures retained the power to prescribe courses of study, lambasting STI laws only for prescribing *how* STI must be studied—with a minimum number of lessons, textbooks in the hands of pupils, and so on. Others condemned "all government control"—in "the election of subjects to be taught," not just "the methods to be used"—and convicted Scientific Temperance on two counts. "A school programme made by a legislature is about as reasonable as a dictionary or encyclopedia edited by the same high authority," quipped an anti-STI editorial in the *Educational Review*, Nicholas Butler's sounding board. Only expert control—cool, neutral, and sober-minded—could block an endless avalanche of lay "partisans" and "agitators," all seeking to impose their own narrow agendas on the schools.[16]

From this viewpoint female activists, especially, lacked the powers of reason and restraint that modern reform required. Mary Hunt and the WCTU were "enthusiasts" and "sickly sentimentalists," schoolmen charged, guided by "mere sentiment and feeling" rather than "intense thought and study." Here educators echoed enemy scientists like Wilbur Atwater, who corresponded regularly with Butler and other school leaders. "They are based too much upon emotion, too little upon reason," Atwater wrote of Hunt and her aides. "Many of the ladies have hearts warm enough, but their heads are not cool enough." To be sure, a wide range of female reformers entered schools during these years to press for medical inspection, vocational training, and other measures. Yet these activists generally cooperated with school-men and condemned STI, just as the rising generation of female school officials did. "As a mother and a school trustee both, I insist that no one except an expert shall interfere with our school curriculum," Illinois native Jesse Bolte told the NEA's Chicago roundtable on STI, endorsing educators' attacks upon it. "What right has any person, or body of persons, to dictate how, when, and where the child shall study any subject without a scientific knowledge of the child's mental status?"[17]

The Child-Study Challenge

Here Bolte cited psychologist G. Stanley Hall, founder and chief publicist of the child-study movement that seized American school reform in the 1890s. From Johann Herbart and other European educators, Hall borrowed the notion that curricula should reflect the natural virtues and especially the natural development of the student. Like the "muscular Christianity" that was sweeping American universities, however, child-study celebrated a par-ticular set of "manly" virtues—strength, energy, and courage—that schools had supposedly stifled by harping upon sin.[18] Scientific Temperance epito-mized "the training of children by wrong methods," Hall wrote, because it appeals "chiefly to fear" rather than "to manliness and the moral nature" within each child. Another schoolman complained, "The trembling hand, the thick speech, the dull senses, the poisoned blood . . . the poverty, crime, and misery of the drunkard are hysterically held up to the gaze of the chil-dren, but the steady hand, the distinct speech, the quick senses . . . the success and happiness of the temperate man are scarcely mentioned." Worst of all, STI embodied the old "Socratic philosophy" that an understanding of evil would guard against it. "Knowledge is a matter of intellect," a third critic explained, "and it may utterly fail to touch the heart."[19]

Stressing emotion over reason, this critique evoked the romantic enthusi-asm of Charles G. Finney and other antebellum revivalists. Indeed, Hall's own efforts to promote child-study drew explicitly upon evangelical images. ("Unto you is born this day a new Department of Child Study," he told an NEA group in 1894—gathered, appropriately, in a church.) But he yoked this venerable romanticism to the more recent science of evolution, arguing that each child's development followed the larger stages (or "culture ep-ochs") of human history. The "juvenile," or early school, years recapitulated a "tribal," or "savage," era, when "reason [was] still very undeveloped" and emotion reigned. Hence STI's emphases upon fact and intellect were not simply naive or misguided; they were "pedagogically wrong and vicious," schoolmen charged, a repudiation of "the God-given faculties with which each child is endowed." Whereas Hunt and the WCTU were too "passionate" in their politics, then, the curriculum they promoted was not passionate enough. Reasoned citizens deferred to educational experts, who alone understood that reason mattered little.[20]

During STI's first decade, only a minority of schoolmen had subscribed to this view. Now that men like Hall gave it their "scientific" imprimatur, however, it spread rapidly. Even in smaller school districts, freshly trained superintendents increasingly embraced the child-study critique of Scientific Temperance.[21] "No one is intemperate because he lacks knowledge, but because he lacks self-control," wrote Frank Perlin, superintendent of schools in Quincy, Massachusetts. "Our attention should be given to the building up of character, self-mastery, and self-respect." Such qualities would never emerge from an education in "wrong-doing and low-thinking," a second small-town schoolman added. Children must study examples of "high moral and religious ideals," he maintained, not drawings of drunkards' stomachs and livers. "Temperance should be taught positively, not negatively," wrote Anna D. Pollard, principal of schools in Plantsville, Connecticut. "It is by studying the Christ life that we become Christ like, not by contemplating a character having opposite qualities."[22]

Protestant Liberalism and Scientific Temperance

As Pollard's remark illustrates, a distinctly liberal theology lay beneath educators' attacks on STI. It too had roots deep in the antebellum era, when many middle-class Americans substituted a gentle Jesus for the angry God of their forefathers. Softening traditional notions of innate depravity, some preachers even imagined a day when evil itself would disappear. Others were less sanguine, cautioning that humans would always be sinful in the eyes of God. But both camps agreed that worshippers could will their own salvation. Those who failed, of course, would burn for eternity: as the saying went, there was no middle ground between Heaven and Hell. Prophets of the millennium kept one foot squarely planted in Revelation, where angels poured out vials of wrath and the wicked were cast into fire.[23]

After the Civil War, however, a new generation of liberal ministers began to eschew this lonely struggle with sin. Preaching what became known as the "Social Gospel," they called upon worshippers to look outward—into the slums, schools, and sweatshops of industrial America—rather than inward; salvation lay in the streets, not in the soul. Biblical passages about grace and damnation were allegories, ministers insisted, not literal accounts. Here liberal clergymen could draw upon the new techniques of "higher criticism," which promised to render "scientific" interpretations of Scripture. Like child-study, higher criticism was imported from German academia; it celebrated the innocence rather than the iniquities of Nature; but it also claimed exclusive authority to decipher her wonders. In antebellum America, Ferenc Szasz notes, "personal and literal reading of Scripture seemed to form part of the democratic framework of the republic." Now experts would have to explain it, providing the nuance and context that lay readers lacked. Just as child-study suggested that parents did not really understand their children, then, higher criticism "seemed to imply that the average person was no longer competent to read the Bible."[24]

To ministers imbued with this new ethos, Scientific Temperance conjured the same fearsome sermons as the Puritan gospel of their youth. Its leading clerical critic was Lyman Abbott, editor of *Outlook* and the foremost popularizer of liberal theology in turn-of-the-century America. As with orthodox Christianity, Abbott argued, STI's rigid admonitions would spawn cynicism rather than rectitude. "The child builds his faith upon confidence in an infallible Book," Abbott wrote in one of his many anti-STI editorials. "He goes to school and learns that the world was not made in six days, six thousand years ago. His whole system of faith topples over, and he has to reconstruct it from the foundation; perhaps he may never reconstruct it." Similarly, children who found that alcohol was not poison would be lost forever to any temperance appeal. Best, then, to heed the advice of "expert educators," who alone comprehended "vital principles of sound pedagogy"—and condemned STI. So did physiologists, Abbott added, "men of recognized scientific authority" who understood the true effects of drink upon the stomach, liver, and heart.[25]

Experts United

Abbott's remarks signaled a new development in the history of Scientific Temperance: the unity of its expert foes, across different disciplines and specialties. Schoolmen teamed up with liberal ministers to denounce STI's methods, while both groups invoked laboratory scientists to discredit its message. "[T]he attempt to indoctrinate the minds of the youth of the country . . . has been overdone," declared W. R. Huntingdon, pastor of New York's Grace Church and a member of the Committee of Fifty. "Let us not palm off upon their innocent minds pseudo-chemistry and inaccurate physiology." Here Huntingdon alluded to the Committee-sponsored research of Wilbur Atwater, who was lionized by leading clergymen across the country. Abbott celebrated Atwater's experiments on the pages of *Outlook*, maintaining that his calorimeter exposed STI as "scientifically erroneous." After Mary Hunt assembled a handful of researchers to vouch for STI, though, Abbott backpedaled from his physiological objections—and escalated his pedagogical ones. True, he admitted, "experts are not wholly agreed" about the effects of alcohol on the human system. But they did concur about the effects of STI on the child psyche. "Such emphasis upon what is abnormal and pathological [is] harmful and morbid," Abbott concluded. "[E]ducational leaders are united in their judgment."[26]

For their own part, "educational leaders" were only too happy to enlist theologians and physiologists in the anti-STI cause. Although a few naive ministers did support Scientific Temperance, one New York schoolman wrote, "[t]he more intelligent clergy are not quite so easily taken in by 'chaff.'" Educators like Nicholas Murray Butler coordinated with scientists, as well, even inviting Wilbur Atwater to address the NEA's convention in 1900. Like Abbott, Butler recognized that Atwater's studies had not resolved the "controversy over technical points"—namely, whether alcohol was a food or a poison. So he encouraged Atwater to focus upon "the educational and ethical aspects of the subject," even though the chemist could not claim any particular expertise in these realms. "Lord Robert's success in South Africa seems to lie in making a series of flank movements until the enemy is worn out," Massachusetts schoolman Samuel Dutton told Atwater. "I hope some such thing may be done in this case."[27]

With the help of scientists, Dutton noted, educators were carrying the STI struggle back to its original arena: state legislatures. By the eve of Atwater's 1899 experiments, in fact, school leaders in four states had already thrown their weight behind bills to weaken or even repeal Scientific Temperance. Within the next six years, at least nine states would witness new challenges to their STI laws.[28] Testifying before legislative committees, schoolmen emphasized that "physiological wisdom, expert educational leadership, and intelligent sociological zeal" were all arrayed against Scientific Temperance. Even WCTU allies admitted that the nation's "whole educational machinery" opposed STI.[29] So Mary Hunt would throw a wrench into it, hoping to expose cracks in the united front of expert authority.

"Outside Evidence": The Search for Schoolmen

Initially, Hunt tried to rebut educators by citing "her" physiologists— Winfield Hall, August Forel, and the handful of other scientists whom she had recruited. As a Connecticut WCTU worker explained, though, the new expert challenge required a parallel set of expert allies. "Even if we can give the names of the M.D.'s on the Advisory Board . . . it will carry little weight," Minnie P. Cooley wrote. "[I]t is well for us to be able to tell what educators are behind this department." Cooley herself did not care which authorities supported STI, she emphasized. Yet "we need all the aid we can get in influencing public opinion," she wrote, "and that argument often convinces where others fail." Nor did it help that STI's lay defenders were primarily female, Cooley added. "You know there are many people who are more ready to believe what some man says than what any woman says," she noted. To defend STI, then, the WCTU would need to cite "outside evidence"—in the form of male educators.[30]

Actually, Hunt had already lined up several prominent schoolmen behind Scientific Temperance. Atop the list stood U.S. Commissioner of Education William T. Harris, bête noire of G. Stanley Hall and his child-study clique. During a famous debate at the 1895 NEA convention, Harris admitted the importance of children's natural interests but rejected the notion that curricula should revolve around them. Instead, he argued, schools should cultivate the cerebral powers that students would need to master such impulses. Like Mary Hunt, Harris was born in the 1830s to a Connecticut family that descended directly from the early Puritans; again like Hunt, he proudly invoked these ancestors in his paeans to human reason. But reason would not magically emerge from "mental discipline" in any subject, as an earlier generation of educators had supposed. Rather, Harris argued, it required specific, sustained instruction in the rich intellectual traditions that transcended each individual.[31]

From the start, Harris praised STI for disseminating "the results of scientific investigations" about alcohol. If it had proven "difficult to teach," he wrote, the problem lay less in STI's emphasis upon facts than in the "pedagogical form" they assumed. In traditional disciplines, "earlier lessons prepare the pupil step by step for the later lessons," Harris explained. But in new branches like Scientific Temperance, "the necessary step to connect one point with the next are [*sic*] lacking and the instruction can not be made so thorough." As soon as schools established this sequence of "progressive lessons," Harris predicted, STI would take its place alongside arithmetic, grammar, and the other established subjects. It too would require regular textbook instruction, which reminded students of "what has already been accomplished" in the complex world they would inherit.[32]

For at least three reasons, however, Hunt could not find more than a handful of other prominent educators to support STI at the turn of the century. First of all, Hall and child-study were at the peak of their short-lived influence.[33] Schoolmen were hesitant to identify themselves with Harris's knowledge-centered approach, which seemed sterile and antiquated by comparison. Second, those educators who still emphasized content over interest tended to share Harris's fear of "crowding the curriculum" with newfangled subjects. Harris himself made an exception for STI, since it targeted "a great evil, perhaps the greatest evil abroad in the land." To other conservatives, however, Scientific Temperance represented yet another constraint upon their beloved "three Rs." Some critics linked it to "vocational" novelties such as bookkeeping and stenography, which also drew time and resources away from traditional courses. Others dismissed it as a "health fad" like vaccination, textbook fumigation, and medical inspection.[34]

Despite their disagreements over curriculum, finally, leading American educators agreed that experts—not amateurs—should determine it. Dissecting "curriculum ferment" in the 1890s, Herbert Kliebard has shown that "there was not one but several reform movements . . . each with a distinct agenda for action." Yet the various professional "interest groups" that Kliebard identifies shared an antipathy to *lay* interest groups in school affairs, uniting men as ideologically disparate as Hall, Harris, Charles W. Eliot, and Joseph Mayer Rice.[35] For most leading educators, then, STI's popular origins were anathema. Indeed, even its handful of supporters took pains to distance themselves from Mary Hunt and the WCTU. Harris served briefly on Hunt's advisory board, but apparently abandoned it after his endorsement raised eyebrows in academia.[36] Normal school instructor William Mowry helped introduce STI in Boston, defended it at the NEA, and even hired Hunt to speak at his famed summer institute on Martha's Vineyard; yet he opposed textbook requirements and other legal specifications, as befitted a man who elsewhere argued that "professional experts . . . must be placed in charge of the schools." Likewise, Iowa superintendent Henry Sabin praised STI but blasted its lay advocates for diluting educators' authority. If laws could mandate the number of STI textbook pages, he asked satirically, why not page size and printing type? (The women who framed such measures evidently believed that schoolmen "have very little sense, and still less conscience," an angry Sabin added.)[37]

Hunt Answers the Critics

In short, STI failed to amass what Minnie Cooley called "outside evidence"—a coterie of friendly educational experts, prominent enough to sway the public. Although she never abandoned the quest for schoolman support, then, Mary Hunt moved to rebut educators' attacks by herself. In articles, leaflets, and speeches, Hunt turned their child-study rhetoric on it head. Appropriating flashy terms like "culture epoch" and "psychological hour," she tried to show that they argued *for* Scientific Temperance, not against it. Ultimately, Hunt hoped, a shrewd use of expert concepts and theories would compensate for a severe shortage of expert allies.

The first task was to establish that STI's knowledge-based approach fit youngsters' "natural interests," child-study complaints to the contrary. Even the great Stanley Hall admitted that "'one of the most marked characteristics of the child . . . is his desire to know facts,'" as a Hunt aide noted. Scientific Temperance satisfied this instinct by stressing the scientific consequences of drink. Would the child be bored? Not if teachers followed a graded curriculum and the STI recommendation of three lessons per week for fourteen consecutive weeks. If lessons are spread casually across the school year, Hunt wrote, "the pupil's interest is not so well sustained, nor proficiency so easily acquired." Child interest, in short, demanded precisely the type of rigid, sequential lesson format that William T. Harris championed—and G. Stanley Hall loathed.[38]

Similarly, Hunt and her STI allies turned Hall's concept of "culture epochs" to their own ends. "Psychologists tell us that all the stages of development, from the savage to the civilized man, are shown in the progress of the individual from infancy to maturity," Hunt wrote on the pages of her *School Physiology Journal*. From this theory, however, she drew the opposite conclusion from Hall: schools must warn children about alcohol, lest they remain permanently mired in a primitive state. "Boys in the primary grades are little cherubs," Hunt wrote. "When they reach the fourth school year, glimpses of the animal begin to appear, and in the fifth and sixth years of the savage. If faithful guidance and instruction are lacking there is danger that some may be full-fledged Arabs by the time they reach the eighth year." The final remark was satirical, of course, but it reinforced Hunt's larger point: that Scientific Temperance was dictated by "the nature of the children themselves," just as its critics would wish.[39]

Child-study rhetoric also helped rebut STI's conservative opponents, who feared that new courses took too big a chunk out of the school day. To make room for Scientific Temperance, Hunt suggested, teachers should pare down arithmetic and geography, remove grammar from the younger grades, and eliminate special reading courses altogether. Citing the fashionable pedagogic creed of "correlation," she argued that reading should be taught in tandem with other subjects rather than separately.[40] More generally, she joined her child-study critics in heaping scorn upon the traditional three Rs. Scientific Temperance, wrote one supporter, "is far more valuable . . . than a knowledge of just how many were killed in each of the battles of the Revolution, or of how to solve all the impracticable problems in the advanced arithmetic, or how to parse the involved and difficult grammatical constructions in Milton's Paradise Lost." Here STI's defenders again echoed G. Stanley Hall, who called upon schools to transcend "the fetichism of the alphabet, of the multiplication table, of grammars, of scales, and of bibliolatry."[41]

Of course, their respective rationales could not have been more different. Hunt attacked the three Rs to gain space for Scientific Temperance, which could survive only by displacing one or more of the established subjects. Hall, by contrast, attacked the traditional curriculum for its overweening emphasis upon knowledge—a critique he also applied to STI. Ultimately, then, Mary Hunt and her aides would have to mount a full defense of fact, intellect, and reason in American education. The search for knowledge, they would argue, did not simply reflect children's natural interests; it was also the key to their behavior. "[K]nowledge must always preceed [*sic*] the formation of habit," wrote one STI supporter. "Our educators might with great personal advantage consult some of the authorities upon child culture." Here Hunt and her assistants often cited British evolutionist Herbert Spencer, who ranked "physiological knowledge" as the most important study for human "self-preservation." Others summoned the ancient Greeks, especially their classic tributes to the power of reason. "Teach the pupils to know themselves," wrote a Colorado STI supporter, quoting Socrates, "and to know themselves means not only to know their minds but their bodies."[42]

"The Truth Shall Make You Free":
STI and Religious Authority

Yet the very search for "authorities" outside the educational community underscored STI's declining status within it. Especially after 1900, then, Mary Hunt increasingly invoked an even higher judge. If knowledge about the dangers of alcohol harmed children, "there must be something radically wrong in the most authoritative of all teaching," she wrote. For "all through the Bible, the blessings of right-doing are coupled with vivid descriptions of the consequences of following the wrong." After God delivered "the 'Thou shalt nots' of Sinai,'" Hunt added, he ordered the Hebrews to "teach them diligently unto thy children." Yet the Ten Commandments "never drove a man, woman, or child to lie, steal, or break the Sabbath." By warning children about the perils of drink, then, schools simply "follow in method the great Teacher."[43]

Across the country, Hunt's WCTU lieutenants echoed her scriptural defense of Scientific Temperance. "How unfortunate," quipped an Ohio STI worker, "that the Bible is so unpedagogical as to teach the Decalogue as well as the Beatitudes." The prodigal son was hardly "an ideal young man," she noted; yet Jesus himself held this example up to view, "some modern ideas of pedagogy notwithstanding." In Chicago, a WCTU official returned all the way to the Book of Genesis to rebut a school superintendent's charge that STI promoted drink. "We may safely conclude," she declared facetiously, "that the law given to Adam and Eve was responsible for their sin, and indeed that the whole Bible was a mistake." Scripture also helped New York's WCTU president lash back at Cornell physiologist Burton Wilder, who told a 1903 teachers' meeting that strict admonitions against alcohol were unnecessary. "Over and against Dr. Wilder's very liberal views, we might set the words of Christ, 'Watch ye therefore for ye know not when the Master of the house cometh,'" argued Frances Graham. "In other words, be on guard always."[44]

Of course, liberals could quote Scripture as well. Wilder restricted himself to the Bible's injunctions about wine, which "'maketh light the heart of man.'" Blasting STI's claim that fact impelled action, educational critics cited Jesus' instruction, "He that doeth the will shall know of the doctrine." Both camps noted that "the truth shall make you free," perhaps the most quoted scriptural excerpt in turn-of-the-century America. Yet as Walter Lippmann famously remarked with respect to this passage, "the truth is a thousand truths which grow and change." Rather than settling the score, then, the passage simply raised anew Pilate's ancient question: What is truth?[45]

In modern America, Mary Hunt realized, "truth" was often what specialists construed it to be. Rather than simply citing Scripture, then, Hunt also sought "pastors of all denominations" who would endorse her interpretation of it. Her best-known supporter was Boston preacher Joseph Cook, whose famous "Monday Lectures" had inspired Hunt's original foray into temperance work. Other minister allies included Charles L. Thompson, a prominent New York Presbyterian; Francis Edward Clark, founder of the Christian

Endeavor youth society; and A. C. Dixon, who would later edit a famous series of books called *The Fundamentals*. As Dixon's title suggested, these men shared a commitment to traditional theology and an impatience with so-called higher criticism. But they also took a deep interest in the sciences, arguing that biblical and modern knowledge reinforced each other. So they meshed well with STI, which likewise combined a strong devotion to science with a rigid scriptural literalism. Together, Hunt boasted, these pastors represented "the best men" in their respective denominations.[46]

As in her struggle against schoolmen, however, the most influential ministers generally opposed Scientific Temperance. Hunt's allies had no mouthpiece to match Lyman Abbott's *Outlook*, for example, which reached 100,000 readers by 1900. In Massachusetts alone, STI's enemies included Francis Peabody, a Harvard minister and member of the Committee of Fifty; William Lawrence, the state's Episcopal bishop; and George Hodges, Lawrence's successor as dean of his church's Theology School. Scientific Temperance was also routinely denounced by prominent Unitarian clergymen (who charged that STI forced children to study "subjects beyond their comprehension") and by the *Congregationalist*, Massachusetts's leading religious newspaper. Foes like Peabody coordinated closely with Hunt's scientific critics; scientists, in turn, ridiculed her ministerial allies. Cook, especially, was a common target for scorn and satire. "His favourite method of dealing with a scientific writer is to quote from him all sorts of detached statements . . . without the slightest regard to the writer's general system of opinions," essayist John Fiske wrote of Cook in 1880. "The 'latest scientific intelligence,' with him, means the last book or article which he has glanced over without comprehending its import."[47] Fiske reprinted the attack in 1900, because the method it criticized "is a favourite one with other controversialists than Mr. Cook"—including, presumably, Mary Hunt. That year, Hunt found herself embroiled in a bitter struggle over Massachusetts's STI law. On one side stood almost the entire state intellectual establishment, ranging from scientists to educators to ministers, who wished to limit Massachusetts's STI requirement to only two school years. On the other stood Mary Hunt, who hoped to strengthen the law by adding penalties and text specifications. Hunt did manage to enlist several lesser-known pastors like Charles L. Morgan, the minister who submitted "her" bill to the legislature. Since these conservative clergymen lacked the influence that her enemies commanded, however, Mary Hunt would have to manufacture it. Galvanizing the pastors into a grass-roots army, she would parry the expert challenge—and contradict her own paeans to expertise.[48]

"Raising the People"

The first battle occurred in New York, which Hunt had always regarded as the most "strategic point" in her movement. School boards easily evaded its 1884 STI law, which merely mandated that they make "provisions" for Scientific Temperance.[49] So Hunt demanded a stiffer statute in 1895, provoking a storm of protest from New York's educational leaders. Columbia president Seth Low, state university chancellor Anson Upson, and state superintendent Charles Skinner all condemned Hunt's bill as a "meddlesome interference with educational freedom," because it mandated textbooks, a minimum number of lessons, and penalties for noncompliance. Skinner sent an anti-STI questionnaire to prominent school officials across the country, eventually publishing dozens of their responses in an aptly named book, *Protest against the Bill.* Hunt's 1895 measure also drew fire from the Public Education Association [PEA], a coalition of wealthy New York women that formed the same year to help promote school leaders' agenda.[50]

Given the opposition of influential figures like the PEA's Mariana Griswold van Rensselaer, Hunt could no longer plausibly claim that America's "mother-interest" stood united behind STI. Indeed, as Skinner's remarkable *Protest* manual indicated, America's *education* interests stood virtually united *against* it. Growing numbers of female school officials joined their male counterparts in blasting Hunt's bill: in the midst of the STI struggle, for example, Rensselaer would be appointed a school inspector by New York mayor William Strong.[51] Deprived of its former claim as the nation's female standard-bearer, the WCTU could not fight this foe alone. To sustain Scientific Temperance, Mary Hunt would need a new popular movement that drew—as her adversaries did—on citizens of both sexes.

She seized upon the ministry, forming a statewide "Central Committee" of pastors to coordinate with the WCTU—and to exert local pressure upon politicians. "[W]hile resolutions passed by State societies have a value," declared a committee circular to New York clergymen, "the wishes of no class are so influential with a legislator as those of his own constituency." Attaching two copies of a petition in support of Hunt's bill, the circular asked every minister to read it to his congregation, request "a rising vote of the audience in favor," and record that vote on each copy—one for the house of representatives and the other for the senate. Pastors were then instructed to sign both documents and forward them to Mary Hunt, who took up residence in Albany for most of 1895. "[E]very moment . . . has been occupied, speaking three times a day," she wrote an associate, "raising the people to defend their

new temperance education laws." Seeking to "focalize sentiment" upon critical legislators, Hunt's strategy almost perfectly duplicated her Michigan Plan
of the previous decade. Now she married the old design to New York's dense
network of churches, which provided a natural venue for local political action. In some cases, the marriage was literal as well as figurative: while
Brooklyn's Mary Newton served as state WCTU superintendent for Scientific Temperance, her husband Albert sat on the Central Committee.[52]

The results were astounding, even to a devout optimist like Mary Hunt.
Despite the steadfast opposition of Skinner and other schoolmen, Hunt's
strengthened law sailed unanimously through the New York legislature. The
measure then went to the desk of Republican governor Levi Morton, a former vice president and an aspirant for the White House in 1896. Canceling
a scheduled trip to Europe, Hunt spearheaded a fresh campaign to persuade
him to sign the measure. At Morton's request, meanwhile, Skinner collected
names of prominent educators who would support a veto. The governor
signed the measure on the last possible day, causing a bruised Skinner to
complain that Morton had double-crossed him. "He gave me the assurance a
long time ago that he would not approve this bill, and asked me to give him
educational backing," Skinner wrote a friend. "I think I did this to his heart's
content." In his memorandum approving the new measure, however, Morton
explicitly rejected the theory that educational experts should determine curricula. "The power to establish school implies the power to prescribe courses
of study," Morton wrote, "and what these courses of study shall be must be
largely a matter of legislative policy." Since the new law had passed "without
a dissenting vote in either branch," Morton added, it surely reflected "a widespread public sentiment." The legislature created schools; the people controlled the legislature; so school curricula should reflect "public sentiment."[53]

The Majoritarian Defense

Often quoted by Mary Hunt and her allies, Morton's remarks would provide
the touchstone for a revamped majoritarian defense of Scientific Temperance.[54] After a second victory in New York, where she rebuffed an 1896
challenge to her new measure, Hunt created other committees of ministers
to press for strengthened STI laws in Illinois and Massachusetts.[55] These
pastors and their congregants were not experts, as they readily admitted. But
they were citizens, and that was enough. "[W]hose are the schools, anyway?"
asked Boston minister Albert Plumb. "Do they belong to the hired servants
of the public, or to the people, the fathers and mothers?" Lest anyone miss

the point, Plumb underlined his answer: "The schools are *our* schools." Established of, for, and by the people, public education was also funded by their taxes. For financial as well as philosophical reasons, then, lay citizens retained the right "to decide what their free public schools shall teach," as New York's Central Committee declared.

Professional educators, meanwhile, were enjoined to follow their lead. "These educational officials here have said: 'We are experts. . . . The schools are placed in our hands. We should be left to run them as we think best,'" Plumb wrote. Yet this claim reflected "the old medieval idea that knowledge . . . should be confined to a special guild," he added. Laypeople well understood what they wanted education to do; indeed, they often possessed a commonsense wisdom that professionals sorely lacked. "Educational experts are apt to be so wrapt up in their pedagogical theories," Plumb wrote, "that they are not awake to the great moral and social exigency that is upon us." Their calls for "cool" and "neutral" thinking merely underscored experts' isolation from the real world, he argued. "[I]t is impossible for a true man to be wholly impassive in regard to any truth clearly discerned," he declared. "The very affectation of neutrality is itself a powerful sermon to show the unimportance of the subject at hand."[56]

Here was an attack upon the logic and culture of expertise, not just upon a particular set of experts. Seeking to enlist her own allies, Mary Hunt had implicitly (and sometimes explicitly) endorsed the notion that experts—cool, neutral, and specially trained—should have a special voice in setting school policy. Now she too declared that Scientific Temperance was "a question of majorities," best answered by lay citizens rather than professional educators. Preparing a new set of leaflets for her workers in the field, she once more called upon them to stream into local schools. The leaflets addressed the most common pedagogical objections to STI: that it was too complicated, that it stunted young children, and so on. At root, however, the new campaign had a simpler goal: to remind public educators that the public still ruled. "There is a certain set of men in educational circles [who] seem to think they own the schools," Hunt wrote, "that the people, the parents have no right or voice in what should be taught their children." Marching back into the schools, her female army would prove otherwise.[57]

By this time, more and more local superintendents shared the professional outlook of Nicholas Murray Butler, Charles W. Eliot, and the other doyens of the Educational Trust. So they, too, denounced the new wave of visitors from the Woman's Christian Temperance Union. Schoolmen "dislike very much being interfered with in there [*sic*] elaboration of an educational system," wrote an STI worker from tiny West Galaway, New York. "None of [them] have taken very kindly to the new departure of being watched, questioned, or advised by their constituents." In faraway Arizona Territory, a WCTU visitor reported a similar antipathy to her presence in the schools. "This subject is treated much as though the local superintendents appointed by the W.C.T.U. were invaders," she complained. Yet "as citizens and mothers," the WCTU had the authority—indeed, the duty—to monitor local schoolmen. "American citizens," a third WCTU visitor explained, "do not delegate their rights to their public servants."[58]

Visitors were little fazed by educators' claims of expertise, which they greeted with a mixture of scorn and satire. In Utica, New York, the WCTU's Mary E. Tallman marched into the office of "school superintendent George Griffith, P.H.D. [sic]," and ridiculed his naive predilection for "filling the children with so much good" that "there would be no room for evil." Evil, Tallman responded, "was all about us"; the only issue was whether youngsters should remain "in ignorance of the cause." (Indeed, Tallman hastened to note, Griffith's own son "is not so full of good but he has room for the cigarette.") In Brooklyn, meanwhile, state STI superintendent Mary Newton called for a different kind of professionalism—based not on abstract rules and principles, but upon the wishes of school patrons. "Learned (?) men tell us that [STI] interferes with pedagogical standards and professional recognition," Newton wrote, punctuating her skepticism with a question mark. "But what does the average mother care about professional recognition so long as her boy is trained [in] habits of sobriety?" True, lasting recognition "comes by achieving," Newton added. "Educators will get this by training boys and girls to become sober, industrious citizens. That is what the state wants and what the [New York] law provides for."[59]

Storm Clouds on the Horizon

Here Newton took aim at Brooklyn school superintendent William Maxwell, who was leading a statewide campaign to overturn Hunt's strengthened STI law. Author of a popular school reader, Maxwell was surely "a good authority on grammar," Newton admitted. "[B]ut in regard to teaching physiological temperance in the public schools," she added, "he is entirely incompetent." Indeed, Newton suggested, Maxwell was not an authority on teaching *anything*. "Superintendent Maxwell never taught school a day in his life," she noted, "but has picked up considerable information in relation to school management." That qualified him to purchase desks and

chairs for Brooklyn's classrooms, perhaps, but not to "select suitable studies" for its children.[60]

Earlier that year, Maxwell had circulated a memorandum requiring teachers
to provide fifteen-to-twenty-minute "reading periods" in STI. His directive
clearly complied with the letter of the 1895 law, which mandated textbooks
"in the hands of pupils" and a minimum number of lessons. Coupled with his
omission of STI from the official city curriculum, however, Maxwell's memorandum smacked of a dodge. "Reading does not preclude studying, but it
may not include it," Newton told a WCTU meeting, "and I ask you, mothers
and citizens, how much would you know of arithmetic or geography if you
had been required to merely read a textbook printed on this subject?"

Following Newton to the podium, Brooklyn WCTU president Emma
Pettengill reiterated her support for the new Scientific Temperance law. But
she also defended Maxwell and his allies on the Brooklyn school board,
which had issued its own circular condemning WCTU "interference" in
educational affairs. "I cannot sit here and allow the board of education and
the superintendent of public instruction to be maligned until they have had
a fair chance to show what they will do," Pettengill intoned. Maxwell was
campaigning against the stronger STI law "as a private citizen," she explained, not as a public official. Until the statute changed, he pledged to
execute it. "We are in convention, and should be careful," Pettengill warned.
"It is not our place to criticize beforehand action to be taken in the future."[61]

Such cautionary notes were rarely heard in 1896, the high tide of Scientific Temperance in America. Hunt had just scored her most stunning victory
yet, vanquishing New York's entire educational establishment. More good
news was predicted in Illinois, where a movement for a stronger law had
gathered steam. But storm clouds loomed, already visible in the comments
of Mary Newton and Emma Pettengill. By arguing—erroneously, as it
turned out—that Maxwell had never taught in a classroom, Newton aimed
to discredit his authority over curriculum. At least implicitly, she also acknowledged the authority of those who *had* taught—and presumed that they
would support STI. Yet more and more classroom instructors seemed to
share Maxwell's disdain for the subject. Months earlier, New York's State
Teachers Association had passed a resolution against it; in the next few years,
other teacher organizations would follow suit. Embroiled in their own battle
against turf-conscious superintendents, classroom instructors embraced them
in the struggle against Scientific Temperance.[62]

More ominously, as Pettengill's remarks suggest, members of the Woman's
Christian Temperance Union were also starting to befriend STI's schoolman

foes. Her gentle warnings would be followed by full-scale rebellions in Massachusetts and then Connecticut, where state WCTU officials deferred to liberal educators—and denounced Mary Hunt. In a pragmatic vein, some women claimed that STI simply could not survive against so much expert firepower. But others echoed the liberals' substantive critique of Scientific Temperance, arguing that it robbed children of their natural innocence—and educators of their rightful authority. Even within Hunt's own organization, it appeared, "mothers and citizens" were hardly as united as she liked to think.

5

THE DILEMMA OF MISS JOLLY: SCIENTIFIC TEMPERANCE AND TEACHER PROFESSIONALISM

In 1891, an Ohio school journal sponsored a teachers' roundtable about the state's most recent curricular requirement: Scientific Temperance Instruction. The exchange began with an anonymous letter from a self-proclaimed teetotaler, who nevertheless feared that STI's screeds against alcohol and tobacco would sap parental authority. "[I]t is hard to know just what to say when dear little Johnny looks up in your face and says: 'Why, my papa smokes,'" another contributor acknowledged, suggesting four possible responses that could promote abstinence without obstinance:

"Did you ever ask Papa not to smoke?"

"Perhaps he will not smoke any more, just to please you."

"I don't believe when your papa was a little boy and went to school, that the teacher talked to him about his body and the harm tobacco would do it, and so he learned to smoke, and now it is so hard to stop."

"Do you think your papa would like to have you learn to use tobacco?"[1]

Another respondent noted that her own village had seven saloons, a good barometer of wet sentiment among the citizenry. In accord with her "theory of parental reverence," then, she had initially ignored STI. Yet "at the end," she wrote, "my theory would not fit my conscience . . . [which] upbraided me for not 'obeying the spirit of the law.'" So she began lessons from an old gospel tract, *Primary Temperance Catechism*, which held true to this spirit without sentencing all besotted parents to a pauper's grave.[2]

To Mary Hunt, however, any such compromise was cause for alarm. Especially in immigrant communities, teachers too often tempered Scientific

Temperance to suit the "whims and prejudices of ignorant parents," Hunt complained. "What would [the teacher] do if like opposition met her in the study of arithmetic?" Hunt asked. "If a pupil asserts that two and two are ten, or refused to believe that the world is round, quoting the parent's authority . . . she wisely leads the child on to receive the truth, and in the end the parent himself is influenced by it." Meanwhile, in a different appeal to skeptical instructors, Hunt noted that teachers and the WCTU shared a common enemy: school officials. Invoking their new university credentials, these experts looked with equal disdain upon Hunt's army of women and the rapidly feminizing teaching force. Both groups, she argued, should now join arms to block the administrative juggernaut.[3]

To Hunt's horror, however, the vast majority of instructors sided with their schoolman superiors and against the WCTU. Like the Ohio panelists, teachers first rationalized this posture in the language of localism: parental opinion, not "outside agitation," should set the contours of school policy. Eventually, though, they took refuge in professional prerogative. Across the country, teachers declared that lay citizens on all sides of the question could not comprehend it. Hence tipplers and teetotalers alike should cede Scientific Temperance to classroom teachers, the only Americans specifically prepared to discuss and deliver it.

Yet this philosophy was hardly foisted upon teachers by cunning superintendents, as recent scholars have contended. Like industrial laborers, the story goes, teachers initially strove to sustain "workers' control" against a new breed of manager who sought to seize it. Schoolmen triumphed by promoting their own version of teacher "professionalism," which ultimately alienated instructors from "the community."[4] Scientific Temperance reminds us that "the community"—however narrowly or broadly defined—often lacked the organic unity that this interpretation implies. Simply put, teachers faced demands from *competing* publics as well as from critical schoolmen. In the name of parental rights, opponents pressed instructors to muffle STI; and in the name of sisterly solidarity, Hunt and her assistants pressed them to magnify it. Both claimed to represent "the community"; and both were, in a sense, correct. Hence teachers' safest position was a retreat into the professional citadel, which was not merely "imposed from above" (as one scholar has written) but implanted from below.[5]

Likewise, laypeople often embraced expertise in the hopes of quelling public debate. When teachers first complained of parental antipathy, Mary Hunt argued that professional authority should trump it. "No true teacher can remain in the profession," an STI ally remarked in 1884, "if low-bred opinion must bind [her] to the pillars of Philistine iniquity."[6] As instructors' own antagonism mounted, however, Hunt took the opposite tack: the people—not the experts—should determine school curricula. Of course, this solution raised all the same problems as "community" did. For as Hunt well knew, "the people" were hopelessly divided on STI. So even as she proclaimed a popular mandate, Mary Hunt also promoted a "College of Scientific Temperance" to train expert teachers in the subject—and to place it, she hoped, beyond lay rebuke.

Hardly the victims of a deceitful seduction, then, school patrons and teachers alike had rational, self-interested reasons for consummating their marriage to professionalism. But the union brought considerable risks, as well. During these years, most classroom teachers lacked even a high school degree. Although expertise promised to insulate them from lay pressures, then, it could also enhance the influence of school officials. Indeed, these men often cited teachers' poor credentials as a reason to mute—or ignore—their voices. Similarly, Hunt's drive for a new STI college—and a new "professional temperance teacher"—seemed to *discredit* her WCTU assistants in the field. For if only "trained instructors" could educate America about STI, as Hunt now claimed, why should any teacher or school official listen to a lay visitor?[7] Teacher professionalism, Hunt would discover, was a two-edged sword. A powerful weapon for advancing her agenda, it could just as easily strengthen her enemies—and weaken Scientific Temperance.

The Dilemma of "Miss Jolly": A Parable

In 1892, *Forum* magazine enlisted a peripatetic young pediatrician named Joseph Mayer Rice to investigate the quality of instruction in American public schools. Rice's thirty-six-city tour took him through Baltimore, where he chanced to witness a class in Scientific Temperance. He recorded his observations in an impassioned article, republished as part of a book the following year:

> [I]n answer to the question, "What is the effect of alcohol on the system," I heard a ten-year-old cry out at the top of his voice, and at the rate of a hundred miles an hour, "It-dwarfs-the-body,-mind,-and-soul,-weakens-the-heart-and enfeebles-the-memory."
> "And what are the effects of tobacco?" asked the teacher.
> In answer to this, one boy called off, in rapid succession, more diseases than are known to most physicians.

"What brings on these diseases, excessive or moderate smoking?"
"Moderate smoking," was the prompt reply.

City teachers told Rice they were "at liberty" to conduct experiments, a hallmark of the "activity"-based curriculum he urged. Yet they lacked the competence to undertake such reforms, relying instead on the traditional tedium of rote. The "science of education," Rice concluded, had not yet "found its way into the public schools of Baltimore."[8]

Four years later, Mary Hunt weighed in against this new "science" and its dangerous application to Scientific Temperance. In a series of articles aimed at schoolteachers, she acknowledged that the "old method"—that is, "merely getting and reciting lessons"—was bankrupt. Yet she feared the "swing of the pendulum" to an opposite extreme, where laboratory demonstrations and other devices would dwarf her beloved textbook instruction in the dangers of alcohol. The "new education," she wrote, threatens to "narrow the scholar down to just what his own senses and those of his teacher can discover at first hand." Yet the pupil needed pictures, maps, and charts, "which will tell him what others before him have found." Hence even the finest instructor could never take the place of "good textbooks," which "speak louder than a teacher's word."[9]

Caught between these competing injunctions stood teachers themselves, whose dilemma was captured in a lengthy parable serialized by an anonymous Michigan schoolman. The story's heroine, "Miss Jolly," teaches "temperance physiology" via active techniques like calisthenics, quiz-downs, and classification of animal bones. These innovations pique "Miss Prim," a colleague who condemns Miss Jolly's immodest discussions of anatomy. (Miss Prim especially bridles at the term "leg," preferring the far less prurient "limbacy.") Yet Miss Prim shows only intermittent interest in the matter; she devotes the bulk of her energy to her after-hours sewing circle, which is busily stitching "overcoats for the Zulus." Stronger objections emerge from Miss Prim's friends on the WCTU visiting committee, which stops by Miss Jolly's classroom to examine her STI instruction. The committee complains that Miss Jolly devotes too much attention to "bones and flesh" and not enough to alcohol; it also bemoans the absence of textbooks, reminding her that the law requires them. Miss Jolly responds that she complies with the spirit rather than the letter of the law. "You can't do everything at once in a district school," she argues, "and I think if pupils are well grounded in the study of their bodies, it will do them as much good as it will to be harping continually about temperance."

To test this claim, the committee's "Miss Easy" asks the class to tell her what they have learned about bones. Students trip over each other in the rush to answer. Bones are white, hard, and brittle, an eager girl explains; blood flows into them; and *"whisky makes them weaker."* Miss Easy is satisfied. But Miss Prim is not. Displaying a newfound concern with STI, she asks the children to describe "the scientific constituency of alcohol." A painful silence follows, pierced by one young cad who suggests that "they've gone to 'lection." By the end of the visit, the WCTU committee has realized the errors of its ways. Its chairwoman, "Miss Howler," concedes that

> we women had better stay at home and mind our business, or else come here and learn something, instead of coming down here to tell you how to teach your school. I'm clear converted to your way. Mrs. Progress said I was converted before I came, but this has made me stronger than ever in the faith.

"In fact," the parable concludes, "all but Miss Prim seemed pleased with the result of the visit, and Miss Jolly scored another victory."[10]

STI in the Classroom: Theory and Reality

Like any administrative directive, though, the tale tells us more about what its author thought teachers should do than about what they truly did. Hence it also serves as a parable for American educational history, which typically stops short of the classroom door. Relying on schoolman pronouncements, most scholarship has presumed a rough correspondence between instructional advice and actuality. Yet teaching practice remained remarkably constant during each era of pedagogical change, as several historians have recently reminded us. Pressed to discard their dull regimens of rote, some instructors did reform: even Rice's jaundiced account included a handful of classrooms where games, experiments, and other hands-on activities predominated. As student enrollments skyrocketed, however, the vast majority of turn-of-the-century teachers held fast to the time-honored techniques of drill and recitation—from the textbook. Largely untrained, they still regarded the text as the "ultimate source of wisdom" and their "primary tool" in the classroom.[11]

No subject was free from this rigid tyranny of rote—not even new disciplines like Scientific Temperance Instruction. Indeed, STI was probably more traditionally taught than the standard three Rs, where instructors could at least draw from a rudimentary fund of knowledge. But they often lacked these basics in physiology and hygiene, relying upon whatever text they received. "Teacher repeated the lesson paragraph by paragraph, and pupils repeated it after her phrase by phrase," read a report of a sixth-grade STI class in Oswosso, Michigan. "After this, the pupils were called upon to recite the lesson. Each one rising and reciting the whole topic." For those teachers without textbooks, meanwhile, a nearby district prepared thirty questions and answers to recite weekly in every class, irrespective of age. Even in an era of rote, the result must have been horribly humdrum. In a first-grade class, one observer watched the instructor lead her bewildered pupils "in concert" through all thirty passages—in four minutes flat.[12]

Moreover, many of these lessons lacked any mention of the science of alcohol—or of experimentation. According to one Iowa observer, children in an STI class chanted that "the body consists of little balls called cells," that bones are like sticks, muscles like lean meat, nerves like silver threads, and nails like horns. "When the little child returns to his desk," the schoolman continued, "he thinks it curious that he is all balls and still he has also sticks, lean meat, silver threads, horn, etc. beneath this lovely smooth skin." Pressed to demonstrate these concepts, teachers pled a lack of laboratory "apparatus," the schoolman noted. Yet "specimens" were free for the taking everywhere around them, from discarded butchery bones to "some small animal which can be killed." Such articles could help teach the traumatic effects of alcohol as well as the simple truths of anatomy, schoolmen suggested. From Maine to Michigan to New Mexico, they peppered their reports and journals with "temperance experiments." Etherize a frog and "lay bare some of the nerves," one example advised; "excite" the nerves with a knife blade, to show their normal response; then apply alcohol and repeat the process, demonstrating the liquid's anesthetic properties.[13]

Most such experiments did not entail actual vivisection. Deceased animals' bones and organs would suffice: when soaked in alcohol, for example, they established its "hardening" effects. But even these relatively tepid suggestions repelled many female teachers, who simply refused to handle calves' stomachs, pigs' livers, or cats' brains in school. Frustrated schoolmen, in response, added squeamishness to their laundry list of tirades against "feminization" in teaching. "No lady is too timid to prepare a fowl, beef's heart, tongue or even a rabbit for cooking," one educator complained, "[so] why too timid to present these articles to a class? All that is necessary is the courage. . . . 'I will' conquers everything." Yet women teachers rarely mustered this determination, despite another schoolman's patronizing assurances that their repugnance "is only the result of acquired habit" and therefore "easily overcome."[14]

Hence the handful of teachers who did conduct demonstrations typically selected ones that would not require animal displays or dissections. Common favorites included touching a flame to alcohol, pouring alcohol over an egg, or viewing alcohol-and-blood mixtures under a microscope. Still, teachers often lacked the aptitude to explain and apply these experiments. In suggestions to STI instructors, for example, schoolmen stressed that burning alcohol should be used only to "excite the suspicion" of liquor, not to explicate its operation; yet this distinction was lost upon the New York teacher who lit the liquid while telling pupils that it shows "the way alcoholic drinks burn your stomachs," as an outraged observer wrote. The egg experiment, meanwhile, purported to illustrate the coagulating—and, therefore, dehydrating—action of alcohol; in the hands of this same instructor, however, it demonstrated "that alcohol can burn without being set on fire." Finally, the alcohol-and-blood compound allegedly proved that alcohol compressed the corpuscles; lacking a microscope, however, the teacher simply told her class "that when men drink whiskey or other strong drink their corpuscles shrivel up"—a separate and entirely specious claim.[15]

Teachers and the WCTU

Few turn-of-the-century teachers, then, embraced the schoolmen's experimental approach with either the delight or the dexterity of Miss Jolly. Like Miss Prim, many teachers were disgusted by animal experiments; others, no doubt, were deterred by their own inability to explain them. Still others were warned off by WCTU visitors' "insistent emphasis on textbooks," as a New Mexico teacher complained. Botany, zoology, physics, and chemistry "are arranged on a foundation of laboratory work," she noted; yet physiology "is still based on recitation," thanks to the efforts of Mary Hunt and her assistants. While school officials pushed teachers to move beyond the textbook, in other words, Hunt and the WCTU pressured them to stay within it.[16]

From the outset, Hunt presumed that most instructors would side with her, for teachers and the WCTU evoked each other more than either group resembled school superintendents. Gender, of course, represented the most obvious identity between Hunt's "woman's army" and an increasingly female teaching force. Just as important, though, a majority of WCTU members *were* teachers—or had been. At the organization's 1889 national meeting and again in 1902, current and former teachers were requested to rise; in each case, almost the entire convention came to its feet. Handfuls of members had served as principals, superintendents, or board members. But the great plurality ascended only as far as the classroom, where their careers followed the syncopated tempo common to female teachers of the era: they left the schools upon marriage and motherhood, sometimes returning after the deaths of their husbands or departure of their children. This sequence, in turn, reflected the constraints of class as well as the "cult of domesticity." Most female public school teachers, WCTU members included, derived from *"lower* middle class families"—from "the great mass of middling farmers, shopkeepers, artisans, and low-level managers who have occupied the rough edge between manual and intellectual labor," as historian John Rury emphasizes. Gender ideologies notwithstanding, they generally lacked the houseservants and other amenities to sustain continuous classroom careers.[17]

On key educational questions, meanwhile, WCTU women and their school-marm sisters coincided as commonly as their backgrounds and biographies did. Across the country, WCTU conventions endorsed female teachers' demands for equal pay and promotion—and for the rescinding of unequal work rules, especially those banning betrothed women from the profession.[18] In the classroom, moreover, both groups tended to support precisely the sort of traditional pedagogy that their class origins would have predicted. Activity-based techniques reflected the "constructivist" epistemology of their elite proponents; but this contingent view of knowledge—like cognate notions of the "naturally good child"—was often "far slower to penetrate the lower classes," as historian Geraldine Clifford notes. Joining the vast bulk of Americans, most teachers and WCTU women continued to regard truth as singular, textbooks as supreme, and teaching as the simple transmission of one via the other. Claims to the contrary only underscored the ivory-tower isolation of pedagogic "progressionals," one Michigan instructor wrote. "[T]he teacher is surrounded by a hodge-podge of educational pabulum [and] dyspepsia," added an Ohioan. "It has been administered to us by the doctors of education . . . in allopathic doses."[19]

Supervision Versus Support

Actually, academic "doctors" merely prescribed the new bromides. Administration was left to the armies of superintendents who swarmed into turn-of-the-century classrooms, flexing their powers and prerogatives. In urban school systems, especially, burgeoning enrollment multiplied the raw number of supervisors, while bureaucratic centralization magnified their raw might. Teachers squirmed under the close scrutiny of these confident new schoolmen, who demanded absolute obedience to their dictates but simultaneously condemned classroom instructors for lacking initiative and imagination. In rural districts, meanwhile, poor roads and scattered schools prevented super-intendents from visiting each classroom more than once a term. Yet even such officers were often suffused with the spirit of new methods, which they urged upon teachers with heated ardor—and humiliating results. In a typical anecdote, one Michigan teacher recounted how a visiting schoolman took over her class when she failed to execute his brand of "object" instruction. Yet the superintendent fared little better, concluding that the class was "un-able to think" after so many months in her charge. "Of course you feel hurt," the teacher wrote. "If I were a commissioner . . . I would not challenge the teacher's methods, but rather observe the zeal with which she pushed those methods."[20]

Recognizing teachers' strong aversion to this style of supervision, Mary Hunt and her assistants tried to turn it on its head. In rural areas, where schoolmen rarely visited teachers, WCTU women would visit them regularly; and where schoolmen censured teachers, WCTU women would support them. "[I]f the county superintendent does little or nothing," wrote a Min-nesota WCTU leader, "he cannot prevent your doing a great deal. He visits your school once or twice a term. You can visit it every week. He sees the teacher officially, and gives, perhaps, a few very general directions. . . . You can see the teacher often, win her confidence, make her intelligently familiar with our idea of temperance instruction." To prepare WCTU women for such encounters, Hunt designed and distributed "visitation blanks" that com-pared STI to arithmetic and geography along three dimensions: textbook use, specified time, and examinations. The checklist reflected her insistence that STI be taught "as a regular branch," in the same fashion and frequency as any other subject. Yet the blank could also undermine this goal by alienating teachers, who might associate it with the hated "rating sheets" that schoolmen used during *their* visits. Thus Hunt implored STI workers to hide their blanks from the anxious eyes of the instructors they observed. "[H]ave the questions in mind when you go into the physiology class," she advised, "and fill it out afterward."[21]

Similar caution marked Hunt's other suggestions regarding classroom pol-itics. In general, she told a Vermont woman, "great tact should be used in making inquiries, that teachers may not be offended." Theoretically, of course, "any one has a right to visit the schools and find out what work is being done." Yet "there are a great many different ways of doing this," Hunt warned, "and each union should send its most able woman." Preferably, the emissary should be "someone familiar with school work"—hardly a tall order, given the prevalence of teachers in the WCTU. Such women would be more likely to display a "general interest" in educational affairs, laying the foundation for successful STI appeals at some later date. Too often, Hunt noted, WCTU workers asked only about temperance instruction. Yet this type of "direct attack" usually fell wide of its mark, alienating more teachers than it attracted. Better to cultivate an overall spirit of cordiality via teas, parlor meetings, and other informal avenues. In Henry County, Virginia, for example, WCTU women sent teachers flowers and ice cream on the sweltering morning of the state certification test; in Lowell, Massachusetts, they served "light refreshments" at the local high school; and in Worthington, Minnesota, a WCTU ex-teacher hosted an annual party for her former colleagues. More than school visits, Hunt wrote, such events would make teachers friendly to Scientific Temperance.[22]

A Helping Hand?

Yet a congenial instructor was not necessarily a competent one, as countless WCTU observers complained. True, thirty states required teachers to pass an STI examination for certification. Thousands of young initiates spent their winter months busily "cramming" for this test and their spring months anxiously awaiting its results. Yet many teachers who passed the exam promptly forgot what they had memorized, while failing candidates often received certificates anyway. "As many of our teachers have little or no knowledge of the subject," wrote New Hampshire's state superintendent in 1884, "it may be necessary to be indulgent the present year." In Mississippi, likewise, state educators tested all candidates but candidly declared that "a want of knowledge . . . is not a bar to a license."[23]

Surely, though, it was a bar to successful instruction. Thus WCTU visitors sometimes took over STI classes, sharing the material they had gathered from local conventions and Hunt's communiqués. Unlike visiting schoolmen, WCTU women typically lived in the communities they canvassed; so they could teach there on a regular basis, as in the Minnesota school where an STI worker lectured weekly. Yet such "temperance talks" represented a stopgap solution at best. With reliance upon guest teachers, STI would never reach parity with other school subjects: as Hunt often noted, a "regular branch" required regular instruction. Moreover, a WCTU lecture threatened to alienate classroom teachers by underscoring their ignorance. Taking pains to distinguish themselves from meddlesome schoolmen, visiting WCTU women stressed that they taught classes only upon invitation of the instructor—and "in a spirit of helpfulness rather than criticism," as Hunt wrote. Yet even friendly teachers must have felt pangs of inadequacy when they ceded the chalkboard to a WCTU newcomer. "Helpfulness" *implied* criticism, Hunt's disclaimers notwithstanding.[24]

To educate teachers, then, Hunt emphasized a far less intrusive form of assistance: literature. Besieged by censorious supervisors, teachers might easily bridle at yet another visitor in the back—or the front—of their classrooms. Yet they could hardly protest printed information, which could alleviate ignorance without calling attention to it. In Kansas, for example, WCTU women sent a forty-page STI pamphlet to all 11,500 public school teachers in the state; in Iowa, they distributed similar tracts to 3,900 such instructors. Yet the centerpiece of Hunt's teacher education effort, the *School Physiology Journal*, never achieved this type of ubiquity in American schools. Published monthly from her Massachusetts headquarters, the *Journal* included teaching suggestions, updates on scientific research, and "authoritative quotations" regarding the harms of alcohol. At sixty cents per year, however, it was too costly for most unions to provide to all local teachers. Hence unions urged school boards to purchase the *Journal;* failing that, they pressed teachers to subscribe themselves.[25]

Neither effort bore significant fruit. In 1897, for example, just 1,321 of New York's 35,000 teachers received Hunt's periodical. Its popularity was greatest in Pennsylvania, where total subscriptions topped 1,800 in 1896; but they fell to less than half that the following year, after the Philadelphia school board canceled its several hundred orders. Turning to teachers, Hunt found them even less willing to purchase the *Journal*. "We realize that teachers have many demands upon their purses . . . and that many of them do not feel like making any further inroads upon their incomes," she told a Nebraska STI worker. "Yet if they could only be made to feel a real enthusiasm for the subject they might be willing to subscribe."[26]

Teacher Resistance I: The Burden and the Boredom

Rarely, though, did teachers develop such "enthusiasm" toward Scientific Temperance. In Hunt's analysis, resistance and ignorance made a vicious circle: teachers lacked the knowledge to develop interest and the interest to develop knowledge.[27] Yet their repugnance actually had deeper roots, embedded in their overall distaste for Gilded Age curricular differentiation. "I cannot crowd this temperance teaching into my school," complained one teacher, in an 1887 letter to her local WCTU. "Every year something is added to the requirements for the next grade; but the effects of alcohol is one straw too many." Saddled with drawing, music, manual training, and a host of other novelties, teachers were in no mood to add yet another subject to their load.[28]

Many school boards compounded the problem by omitting STI from their official courses of study, forcing teachers to frame their own provisions for it. This proved especially problematic in rural schools, where students of wide ages and abilities were often seated in a single room—and "daily recitations . . . were already more numerous than was desirable," as an Iowa county schoolman noted. Even in graded buildings, however, teachers groaned. "I cannot find that the grade of school makes any difference," reported a Brooklyn WCTU woman, after surveying city teachers about STI. "The majority . . . are opposed on account of the time it takes. The teachers are overtaxed, and the children too."[29]

Indeed, teachers often cited student tedium in their salvoes against Scientific Temperance. In Perth, New York, a teacher grumbled that STI bored her pupils—and "if she were taught so much 'hum drum' all the time it would make her drunk." Others argued that STI did exactly that, not by stultifying interest but by sparking it. "So much talk about [alcohol] causes pupils to try it," wrote one teacher; another reported that STI had led her students to "experiment" with distillation. More than anything else, though, teachers complained that their pupils were confused. Especially in the primary grades, wrote a Massachusetts observer, children "were found to be repeating accurate statements . . . the meaning of which was utterly beyond their grasp." By the time they could comprehend such claims, students were too stupefied to care. At "the bare mention of the word physiology," wrote one Illinois teacher, seventh and eighth graders "assumed an air of utter indifference, if not disgust."[30]

Teacher Resistance II: The Wet Threat

Yet the sharpest opposition came from parents, as one Maryland principal told a WCTU visitor. Across America, "hundreds of thousands of mothers . . . do not want such instruction given," he cautioned. Although not as organized or persistent as the WCTU, these women also visited classrooms to voice their sentiments. "I called to arsk you, Miss, not to give 'Arrietann any more lessons about people's insides," complained an immigrant mother in a 1902 cartoon, confronting her daughter's befuddled teacher. "She don't at all like it, and besides, it's rude!" Others objected more specifically to

STI's screeds against alcohol, which parents claimed to use without incident or injury. Across the country, they took their case to teachers, principals, and school boards—or sat on the boards themselves, as in the Minnesota town where beer-drinking Germans snubbed STI and the "vimens" who supported it.[31]

In most districts, meanwhile, these boards still retained the discretion to dismiss teachers without reason or recourse. Eventually, superintendents would wrest much of this appointive power away from lay boards. Before 1900, however, city and country teachers usually served at the whim of locally elected school directors. In wet areas, then, prudence as well as principle prompted teachers to temper their zeal for Scientific Temperance. "[T]eachers who are earnest and emphatic in urging this subject . . . are apt to come in collision with members of the board and lose their situation," a Philadelphian reported. In Brooklyn, one particularly hostile director—"the brewer Scharmann"—threatened to fire any employee "who should dare to . . . say a word to a pupil against beer."[32]

Since drink cultures cut across ethnicity and region, teachers discovered, no district was safe from such risks. Instructors in tiny Riverhead, New York, "daren't talk" about STI, a WCTU woman wrote, as "several cigar manufacturers . . . are on the Board of Education and their children attend school." At an Illinois teachers institute, a friendly schoolman blocked debate on several STI resolutions for fear that they would be reported in the county newspaper, "and teachers voting for them might lose their schools in places bitterly opposed to temperance." No wonder, then, that instructors in such areas limited their classes to casual "cold-water" songs—or simply "let the whole thing go by default," as a Wisconsin educator observed. Others delayed STI until the spring, when student attendance—and school board oversight—tended to subside.[33]

Teacher Resistance III: The Siren of Professionalism

On this sea of controversy, professionalism shone as teachers' safest harbor. Increasingly, they condemned lay directives of every stripe for diluting "the discretion of the teacher," as several New Yorkers told a WCTU visitor. Classroom instructors "are perfectly competent to do their work in an intelligent manner," an Ohioan added, "without any law to tell them exactly when, how often, and under what circumstances they shall act." Here he took aim at new STI measures requiring textbook instruction, pupil examinations, and a minimum number of lessons. Yet teachers reserved special ire for penalty clauses, which nineteen states added by 1905. Fewer than half of these states specified that recalcitrant instructors would lose licenses or jobs; most of them targeted school boards, which risked forfeiture of funds if they neglected STI. To critics newly imbued with a "professional" ethos, however, *any* penalty represented "an insult to the teachers," as one educational association resolved. "A reaction has set in against the mode of teaching temperance enforced by legislation," a New York teacher concluded, "rather than left to the judgment of the instructor."[34]

Teachers with college or normal degrees were most likely to take offense. "[T]hey have been specially trained to teach and they feel that by virtue of that training they should know more about what and how children should be taught than those untrained," explained a WCTU observer in 1910, "and hence, they are apt to resent even slight interference with their methods and work." Even at this late date, only 5 percent of American classroom teachers had any postsecondary education. Significantly, however, untrained instructors adopted the same mantle of professional pique in their salvoes against STI. "Ninety-nine percent of the teachers of New York State are probably thorough-going temperance people," declared an 1895 resolution signed by eighty teachers in Watertown, New York, "but the legislature should no more tell us how we should teach than it should tell the physicians how to prescribe."[35]

Forging their own professional fortresses, schoolmen blocked most such bids for teacher autonomy and authority. Their "One Best System" rewarded obedience rather than independence, as David Tyack and a host of subsequent scholars have reminded us. Yet these very purveyors of "close supervision" blasted Hunt and her aides for diminishing the discretion of classroom teachers. The irony did not escape STI supporters like Magee Pratt, a Connecticut minister. In struggles to centralize his state's school administration, Pratt noted, "every educationalist . . . asserted that the teachers in the rural schools were incompetent, half trained, and unfit to manage the schools unaided." The same educators condemned STI for compromising "teachers' rule," in effect asking the state "to allow these incompetent people to determine its educational policy."[36]

Conversely, teachers who otherwise scoffed at schoolman paeans to classroom "professionalism" seized it during the struggle over STI. In the hands of hostile bureaucrats, the concept easily became a club to enforce system-wide directives: teacher "professionalism" connoted submission to the will and whim of *administrative* professionals, as critics like Margaret Haley realized. Yet Scientific Temperance represented the rare case where schoolmen explicitly endorsed teacher discretion rather than docility. Hence teachers happily embraced their educator enemies in the march against STI, joining hands under the banner of professionalism. Even the teacher journal *School,* chief foe of Nicholas Murray Butler and the other sultans of centralization, praised these officials for resisting the "radical ideas" of Mary Hunt and her "coterie of social reformers."[37]

Hunt's Response I: From Professionalism to Majoritarianism

In the face of this new challenge, Hunt would have to switch course as well. During STI's first decade, she often defended teachers from lay attack by citing their professional skills and prerogatives. After Ohio legislators complained that "few teachers would be competent" to instruct STI, for example, the WCTU's *Union Signal* suggested that the "fraternity of pedagogues" should "demand a most humble apology for this affront." Indeed, the *Signal* joked, "[t]he 'least mite' of a . . . scholar could have legislated more aptly and reasoned more intelligently on this subject than the present general assembly of the Buckeye State."[38] By the mid-1890s, however, teachers often regarded STI as an "affront" to their competence and authority. As in her struggle against enemy educational officials, then, Hunt would have to mount a renewed defense of lay control in school curricula.

Initially, she linked the two groups of dissenters. As "public servants," Hunt declared, school officials and instructors were both beholden to public sentiment. "Is it not equally the duty of the school boards, superintendents, and teachers to enforce the law requiring Scientific Temperance," she asked in October 1894, "not to philosophize about what the law should be?" When New York teachers resolved to resist Hunt's new STI measure the following summer, however, she singled them out for special scorn. "[S]urprising as it may be to find New York educators uniting . . . to defeat a proposed temperance law, it is still more astonishing . . . to find the teachers of the state openly recommending its violation," Hunt wrote. "To say, 'We will not obey this law because we, the minority, do not approve of it,' is anarchy pure and simple, whether emanating from a teachers' association or from a band of red-handed desperadoes."[39]

Here Hunt's critique of teachers' professionalism shaded into an attack upon their patriotism, as well. She rarely employed such innuendo in her exchanges with enemy schoolmen, almost all of whom were native-born Protestants. Classroom teachers, by contrast, hailed from dozens of different countries and cultures. In Worcester, Massachusetts, for example, first- or second-generation

Irish immigrants held more than a third of the public school teaching jobs; in Milwaukee and Cincinnati, a fifth of all instructors came from Germany. By 1908, in fact, roughly half of public school teachers nationwide were immigrants and their offspring. These newcomers provided an easy target for Hunt and her aides, who often attributed teachers' anti-STI sentiment to their supposedly besotted ethnic heritages. "It is painful, no doubt, for a woman teacher who takes her beer—and there are such—to teach what is found in the books," wrote a Hunt ally from Chicago, noting the preponderance of immigrant teachers in the schools. "It is repugnant—it is humiliating—for men superintendents, principals, and teachers to teach one thing and practice another, and there are such." To keep drinkers from the classroom, Hunt had earlier called upon school districts to hire more females. Now she suggested that they hire more native-born instructors—and bar immigrants of both sexes, at least until they demonstrated their temperate ways.[40]

Finally, Hunt wrote, any teacher who persistently resisted STI should be reported to school authorities. Teas, parlor meetings, and other carrots would only influence a subset of dissenting instructors, she admitted. The most stubborn teachers needed the stick of law—specifically, the threat of losing their jobs. In Ohio, for example, WCTU visitors published a list of non-complying schools in the town newspaper. "That had the desired effect," one observer wrote, "as there are teachers who would teach [STI] rather than have their names appear in print." Elsewhere, however, WCTU women worried that such tactics would damage their already brittle relations with local teachers. In one New York village, for example, an STI assistant gave Mary Hunt the names of several obstinate teachers—and asked *Hunt* to contact the local school superintendent, lest these teachers learn that the assistant had informed on them. Hunt generally declined such requests, believing that school officials responded better to "people in their own community" than to "outsiders" like herself.[41]

Hunt's Response II: Back to Professionalism?

Whatever their source, such complaints had little effect. For even in smaller school districts, Hunt realized, schoolmen often shared teachers' distaste for Scientific Temperance. So did many members of the "community," who increasingly challenged WCTU women at board meetings and elections. Just as teachers invoked professionalism as a sanctuary from this debate, then, so would Mary Hunt. Switching gears yet again, she called for a new breed of "expert" instructor who could hold sway above the flux.

The key lay in normal schools and teacher institutes, which had all but abandoned STI. During its first decade, when most leading educators supported the subject, teacher trainees received heavy doses of Scientific Temperance. Normals included it as a "regular study," while institutes often invited WCTU lecturers to address them.[42] By the turn of the century, however, both institutions either downplayed STI or ignored it completely. "Our WCTU has provided a speaker for the Institute several time [*sic*] in the past," one New York woman noted in 1902, "but now they will not allow any thing said by any one but the conductor." If they examined alcohol at all, meanwhile, normal schools emphasized its "moral" or "social" danger. "Alcohol deadens the will, makes . . . us selfish and finally we degenerate," Maryland normal student Nettie Simpson wrote in her physiology notebook in 1903. "We have no right to do as we please with our lives as they affect other lives." Filling just six of the more than one hundred pages in Simpson's notebook, these "Scientific Temperance" lessons omitted almost all mention of the science of alcohol. In New York, by contrast, normals eliminated alcohol instruction of every sort—on the theory that their students had already received a surfeit of the subject in grade school.[43]

To Mary Hunt, this claim simply underscored the poor schooling of the teacher-trainers themselves. When an 1897 institute conductor remarked that "all there is to be taught about [STI] could be taught in a half an hour," Hunt quipped that "*all that gentleman knows on this topic* he could tell in the time he specified." So Hunt began to probe the establishment of a new graduate school, specifically designed to train normal and institute instructors. The idea had surfaced several years earlier, when WCTU president Frances Willard suggested that the new American University in Washington, D.C., endow a "John B. Gough professorship" of temperance. But the "gospel" tradition of Gough—a leading antebellum reformer—was anathema to the "spirit of science," as an 1891 critic explained. "The aim [Willard] wishes to accomplish is an excellent one, but it is not the province of a university . . . to put into its chairs either temperance lecturers, campaign orators, missionaries, or civil service reformers," he wrote. "[A] professor ought not to be an advocate. . . . [H]e ought to investigate and teach his subject with the single purpose of finding out and helping others to find out the truth." To enter modern academia, then, Scientific Temperance would have to stress science over temperance—and study over teaching. Temperance without science connoted "advocacy," the antithesis of scholarly objectivity. Teaching without study connoted amateurism, especially at a time when normal schools were stressing content rather than methodology.[44]

A [Stillborn] College of Scientific Temperance

In 1894, then, Hunt would return to American University with a very different proposal. Instead of a single professorship, she told vice-chancellor Samuel Beiler, AU should create an entire "College of Scientific Temperance" with chairs in biology, chemistry, pathology, and physiology. Beiler seemed willing, so long as the college stressed "original research" as well as "practical pedagogical work" in these disciplines. "We could hardly take up the work on the ordinary level of a *Normal School,* for the ordinary grade of teachers," Beiler wrote. "But the grade of *teachers of teachers* . . . is just the kind of work we want to do." Like AU's slated medical and missionary schools, he emphasized, the STI college would only admit candidates with undergraduate degrees. Faculty would teach them in the laboratory, fusing investigation and instruction in the same way as other graduate schools did.[45]

Yet there remained one "difficulty in our way," Beiler cautioned, "which confronts most institutions in regard to special work: namely, the *money.*" Many American philanthropists were also drinkers, draining the pool of potential patrons for a college devoted to Scientific Temperance. Recently, Beiler noted, one wine-swilling AU donor had revoked his $10,000 gift in order to protest president John F. Hurst's "pronouncements on temperance." Just to get off the ground, of course, the STI college would require an even greater sum than that. In a hastily arranged meeting, Beiler and Hunt agreed that the school should have at least one full professor who would "command the respect of all scholars"; realistically, they "could not hope to get such an [*sic*] one for less than $4000 or $5000 per year." Four or five part-time specialists in biology, chemistry, physiology, and pathology would each cost another $500 to $1000 annually. Hence salaries alone would require a minimal endowment of $150,000. Adding the price of laboratory and library materials, the college could not hope to survive without a total subscription of $250,000.[46]

Meanwhile, Mary Hunt—not American University—would have to raise it. If AU solicited funds for the STI college, Beiler argued, "it would be but one line of our manifold work." The university was already far short of its own enormous subscription target, set at $5 million by its Methodist sponsors. Under outside auspices, moreover, fund-raising "would appear to be more interdenominational" than any AU effort. At Beiler's suggestion, then, Hunt chartered a "Temperance Educational Association" in 1895 for the "creation and maintenance" of the new college. The association was almost an exact replica of Hunt's STI advisory board, which she had established to help stave off attacks from hostile scientists and schoolmen. From education came William T. Harris and William Mowry; from psychology, T. D. Crothers, I. Quimby, and L. D. Mason; and from the ministry, Albert H. Plumb, Joseph Cook, and Daniel Dorchester. These men would retain "power of visitation and investigation of the College." In return, they would have to generate the needed endowment.[47]

In the ensuing months, however, few patrons opened their checkbooks to the new college. Hunt cautioned patience, claiming that a sluggish national economy had kept donations down. But the problem went far deeper than that, touching on the same issue that had plagued STI since the decade began: a paucity of professional allies. Although Hunt did enlist several scientists behind STI, none of them sat on her advisory board or educational association. Crothers, Quimby, and Mason were asylum superintendents, not bench researchers; their specialty lacked both the conceptual rigor and the popular resonance of physiology, chemistry, and other laboratory sciences. Nor did Hunt's minister and schoolman allies represent the best men in their respective professions, as Hunt discovered during her clash with enemy educators. Even Harris was widely regarded as a faded star, paling next to luminaries like G. Stanley Hall.[48]

The Two-Edged Sword

In short, the College of Scientific Temperance was strangled by the same forces that spawned it. Lacking professional supporters, Hunt sought to create a college that would generate them; then the college failed to generate sufficient funds, foundering for lack of professional supporters. After the AU experiment died, Hunt and her allies begged several other young universities to sponsor a program in STI. But all of these efforts also came to naught, further underscoring her weak academic pedigree.[49] Switching gears yet again, Hunt resolved in 1904 to devote her energies "chiefly to the teachers in the classroom" rather than to the preparation of new ones. Given her recent attacks on their skills and knowledge, though, few classroom instructors were inclined to listen. In 1905, representatives from twelve different teacher organizations blasted STI as a blight upon schools—and a slight upon their own professional status.[50]

For these teachers, of course, professionalism posed perils of its own. Like historians today, Mary Hunt assumed that they embraced professionalism—and eschewed "community concerns"—upon the prompting of school officials. Yet teachers had sound self-interested reasons for doing so, as the case of Scientific Temperance clarifies. To privilege "community concerns" risked protests from one set of patrons or the other, each arguing that it represented the dominant sentiment within the boundaries it delimited. Then again, to privilege educational degrees—and other abstract measures of "merit"—left teachers open to the charge that *they* lacked the necessary training for the job. Especially in urban districts, schoolmen would use this new "credential model" to replace insubordinate teachers with more pliant instructors of their choice. For teachers and lay citizens alike, it seemed, the professional sword could cut any number of ways—depending upon whose ox was being gored.[51]

6

"DO-EVERYTHING": SCIENTIFIC TEMPERANCE, THE WCTU, AND LIBERAL EXPERTISE

In 1902, Mary Hunt's own state WCTU leader demanded that she defer to liberal educational experts in the struggle over Scientific Temperance. "[T]he living teacher, the vitally interested teacher, counts for more than do tons of text-books," wrote Katharine Stevenson, in a near-perfect echo of STI's schoolman foes. "Not creed and theory, but life, is what the world of to-day needs; not knowledge, but knowledge vivified; not truth, but truth illumined; not the formal teaching of fact and theory, but the virile teaching of a life." As Stevenson hastened to add, she lacked the skills or qualifications to evaluate such teaching by herself. So she would heed the advice of professional schoolmen, who alone had mastered the mysteries of modern pedagogy. "[T]hey should have the liberty to use what, to their thought, are the best and most advanced methods," Stevenson wrote. "Who, in any state, is so likely to be a judge of these new and better methods as the educators?"[1]

Proud and stubborn, Mary Hunt made a host of WCTU enemies during her twenty-six years atop the Union's Department of Scientific Temperance Instruction. But her antagonists shifted their attacks midway through her career, mirroring larger patterns in American educational politics. Like district schoolmen, Hunt's first foes appealed to local control: every union must govern its own affairs, they maintained, without interference from "outside" WCTU officials. Subsequent critics invoked *professional* control: even in Hunt's native state, Katharine Stevenson proclaimed, STI should reflect expert rather than lay prescriptions. For two decades, of course, Hunt herself had urged Americans to follow the counsel of the "best men," "renowned experts," and "highest authorities" on Scientific Temperance. Now, as in the proverbial Chinese curse, her own wishes were coming true. Across the country, WCTU leaders happily embraced professional pronouncements—*against* STI. Prior enemies had censured Hunt for usurping their rule, insisting that local knowledge gave them rightful aegis over STI. The new cohort embraced *expert* knowledge, which seemed to trump every lay advocate—including Mary Hunt.

Behind both challenges stood Frances Willard, the WCTU's national president and Mary Hunt's perennial nemesis. After splitting with Hunt over partisan politics in the 1880s, Willard had backed the noisy ebullience of the Loyal Temperance Legion over STI's sterile, knowledge-driven approach.[2] As the 1890s dawned, child-study moved this debate onto a fresh, ostensibly scientific plane. Willard seized eagerly upon the new theory, which promised to clothe her old romantic presumptions in the flashy garb of evolutionary biology. After her death, her successors continued to embrace liberal educators—and to denounce Scientific Temperance. Willard's 1898 demise serves as a watershed for most scholars of the WCTU, who claim that it forsook her "Do-Everything" strategy for a narrower focus on alcohol and prohibition.[3] Yet the saga of Scientific Temperance suggests far more continuity between Willard and the leaders she chose to replace her.[4] Both wanted *experts* to "Do-Everything"—while the WCTU would do everything in its power to help.

With the bulk of these professional authorities arrayed against her, meanwhile, Hunt would have to mount a second, more popular appeal within the Woman's Christian Temperance Union. Urging members to resist the Willard clique, Hunt—like her own early foes—stressed each local union's distinctive understanding of its community and schools. Yet she also continued to embrace professional authority, insisting—against all evidence—that "the best men" as well as "the great public" supported STI. Somehow, Hunt believed, her scattered, untrained force of superintendents could sustain these twin imperatives. She was right. Across America, local women spawned a dense, grass-roots "movement within a movement" that challenged enemy school and WCTU leaders alike. Only with Mary Hunt's death would this lay army expire, hastening the demise of Scientific Temperance Instruction in America.

The WCTU and Expertise I: Scientists

As the first set of experts to challenge STI, laboratory scientists were also the first to strike an alliance with Willard and her aides atop the Woman's Christian Temperance Union. The conduit was Wesleyan chemist Wilbur O. Atwater, whom Willard met initially at an 1891 reception for Mary Hunt. Praising Hunt as a latter-day "temperance apostle," Willard dismissed Atwater's charge that STI taught falsehoods about alcohol. With the first reports of his elaborate calorimeter, however, Willard and her allies started to waver. "Professor Atwater . . . made a very pronounced attack upon the school physiologies," Katharine Stevenson wrote Hunt, after encountering Atwater at the 1895 Chautauqua conference. "[H]e intimated that he should publish very soon a refutation of what was taught." Nor could Stevenson "contradict his assertions," she added, since she "had no authority to give save that which he had pronounced unreliable." Here Stevenson alluded to the three medical "inebriests" who sat on Hunt's advisory board, none of them famous for their achievements in the laboratory. Unless Hunt added "at least three well-known scientists" to her board, Stevenson warned, Americans would rightly suspect the accuracy of Scientific Temperance.[5]

At the time, Stevenson served as corresponding secretary of the national WCTU. So she worked closely with Frances Willard, who shared her doubts about STI. "I have never felt so impressed with the weakness of our departmental system as in this matter," Stevenson wrote Willard, after Hunt refused to name any more advisors. "It gives to one woman absolute power. She may appoint anyone whom she pleases, and may carry on a propaganda which is diametrically opposed to the truth." Echoing Stevenson's concern, Willard decreed that all unaffiliated "co-workers" would henceforth require approval from the WCTU's national office. Yet this solution posed its own problem, as Stevenson quickly realized. "[H]ow much time do we have in our busy year and our busy days to investigate?" she asked. "How much do we know of the Advisory Board which stands back of the department of Scientific Temperance Instruction? Absolutely nothing." The dilemma became more acute the following spring, when Hunt released statements by Nathan S. Davis and other prominent physicians in praise of STI. To evaluate Hunt's expert allies, Stevenson told Willard, they would need an expert of their own.[6]

They found a ready one in Wilbur Atwater, who assured Stevenson—and, through her, Frances Willard—that Hunt's supporters were out of step with modern science. "There is a large number of people who have felt it necessary to maintain that the world was created in six times twenty four hours," Atwater wrote Stevenson in 1897, "the consensus of modern geology" notwithstanding. For the moment, he admitted, physiologists lacked the same degree of agreement about alcohol. Thanks to research in his calorimeter and elsewhere, however, a "crystallization of scientific opinion" could be expected shortly—against Scientific Temperance. The entire subject "is building a great structure upon a foundation of sand," Atwater warned. "It is not too late for you to shift your building and put it on the rock of actual scientific fact."[7]

Together, then, Atwater and Stevenson began to plot ways to wrest control of STI. Stevenson initiated a correspondence between Atwater and Willard's successor, Lillian Stevens, promising him that the new WCTU president was "a woman of wisdom and tact." Atwater discussed the matter with other scientists on his Committee of Fifty, emphasizing that "the attitude of the Woman's Christian Temperance Union as a whole is *not* such as we had all inferred from . . . the statements of Mrs. Hunt." He also arranged for Stevenson to meet with physiologist Henry P. Bowditch and theologian Francis Peabody, both members of the committee from her native Massachusetts. The following year, Stevenson came home to assume the state's WCTU presidency. The stage was set for an acrimonious battle with another Massachusetts native, Mary Hanchett Hunt.[8]

The WCTU and Expertise II: Educators

Stevenson returned amidst a new controversy over the state's STI law, which simply required the subject "as a regular branch of study" for "all pupils in all schools." Fearing that local districts were evading the measure, Hunt pressed legislators to mandate a minimum number of lessons, texts in the hands of pupils, and a penalty for districts that failed to comply. To critics like Bowditch, however, children were already receiving too much STI. So he prevailed upon state representative James J. Myers to submit a competing bill, which limited STI to just two grades. As Bowditch boasted, the "Myers Bill" was endorsed by nearly every leading scientist and physician in Massachusetts. With Stevenson's arrival, moreover, even the WCTU might be persuaded to back the weaker measure. "[T]he chances are very good of inducing the legislature to take a rational view of this subject," Bowditch wrote. "We have quite a good deal of evidence that all the advocates of temperance . . . do not sustain Mrs. Hunt."[9]

To Bowditch's chagrin, however, Stevenson and other state WCTU officials endorsed Hunt's "strong law"—and opposed Myers. Stalemate ensued, as both bills went down to defeat. "You see there is not much difference in this crowd of people—they are all tarred with the same stick!" an angry Bowditch wrote Atwater, who had assured him of Stevenson's support. "I do not see how we can expect to do anything with them." In an equally furious letter to Atwater, however, Stevenson explained that her support for the "strong" measure stemmed less from its merits than from his misrepresentations. Specifically, Stevenson wrote, Atwater had spread rumors that she drank alcohol—and that she would help him secure STI's repeal in Massachusetts. Neither charge held a kernel of truth: a total abstainer, Stevenson wished to protect educators' discretion over STI rather than to remove the subject altogether. By supporting the Myers bill, however, she risked confirming Atwater's claims in the minds of her new constituents. (Atwater, of course, denied making either remark.) Even though Stevenson despised Hunt's stronger law, then, her scientist allies left her little choice but to back it. Feeling hurt and betrayed, she turned instead to STI's other cadre of expert critics: liberal schoolmen.[10]

Here, again, Stevenson followed in the footsteps of her recently departed mentor. Since the early 1890s, Frances Willard had used the WCTU's Department of Kindergarten Work to promote and publicize the new child-study "science of education," as she called it. Inspired by Continental theorists like Pestalozzi and Froebel, American educators had discerned "fundamental laws of psychology" that governed young minds, Willard declared. So the WCTU would support G. Stanley Hall and other liberal experts, she added, spreading the good news of "child-culture and child development" through its kindergarten department. Upon its founding in the 1880s, the department had established and operated private WCTU kindergartens. But now that expert educators lent kindergartens their imprimatur, rooting this innovation more firmly in the public schools, lay organizations like the WCTU would simply "co-operate with school authorities," as one Willard ally explained.[11]

In Massachusetts, predictably, most of these authorities heaped scorn upon Scientific Temperance. Prominent opponents included Harvard president Charles W. Eliot and state school board chief Alice Freeman Palmer, the former president of Wellesley and an old friend of Frances Willard. Now Stevenson suggested a public STI conference between educators and the WCTU, sparking howls of protest from two very different quarters: scientists and Mary Hunt. "It will be seen that the medical profession is entirely ignored," grumbled Harvard physiologist George W. Fitz, a long-time STI critic, "notwithstanding the fact that the determination of what is truth in the matter of physiology and hygiene . . . must rest largely upon the physicians." Hunt, for her part, warned against any compromises with enemy educators. "[O]pponents expected to overthrow the total abstinence teaching in the public school with the Atwater experiments," she cautioned. "And now the opposers are evidently trying . . . to work through certain schoolmen." Denouncing the proposed Massachusetts summit, Hunt especially ridiculed the alleged "impartiality" of educators who supported it. Indeed, "the very acknowledgment that there is something to discuss" seemed to prove that schoolmen retained "pre-conceived opinions or no-tions" about STI.[12]

Once it became apparent that the meeting would convene, however, Hunt made sure to attend. She even won a place on the Committee of Twelve that the conference selected to design an STI course of study, hop-ing to sway it towards her own endorsed "Schedule and Standard." But the cards were stacked against her, as Hunt quickly saw. The panel's other three WCTU representatives—Stevenson, Mary G. Stuckenberg, and Louise C. Purington—were all "intimately associated" with Frances Willard. Fluent in German, Stuckenberg had lived in Berlin for several years before returning to serve as Willard's superintendent of "temperance and labor"; Purington, a medical doctor, was national director of the "health and hygiene" depart-ment that Willard also developed. Like "Kindergarten Work," these depart-ments functioned primarily as publicity agents for experts in their respective fields. Purington, for example, distributed literature from state health boards and the federal Department of Agriculture to WCTU women in the field. She also helped lobby for the pure food bill alongside the depart-ment's A. C. True and H. W. Wiley, both close associates of Wilbur Atwa-ter—and erstwhile critics of STI.[13]

Exactly as Hunt feared, then, all three WCTU women joined the com-mittee's six educators in demanding a reduction of Scientific Temperance. Rather than yearly textbook teaching, as Hunt suggested, the panel recom-mended oral instruction through the fourth grade and a complete hiatus from STI in the fifth and seventh. In a stinging minority report, Hunt complained that the committee's plan rendered STI a second-class cousin to other academic subjects.[14] Most of all, she worried that other state WCTUs would now follow suit by "getting together" with educational foes. "You seem to be taking the work of the department upon yourself in this most critical time without in the least consulting with the national superinten-dent," Hunt warned Stevenson in January 1900, "when every unwise move in Massachusetts will re-act on the rest of the country."[15]

"Compromise for Harmony": The Attack Goes National

As if to confirm Hunt's fears, Stevenson soon launched a nationwide move-ment to bring STI into accord with "the best authorities" in American edu-cation. In a circular to other state WCTU presidents, she detailed the Massa-chusetts dispute and solicited their "most unprejudiced judgment" about it. Yet Stevenson's inquiries easily betrayed her own biases, leaving little doubt about what sort of answers she sought. If a joint panel of education and temperance officials in their own states had rendered the same recommendations as the Committee of Twelve, she asked the WCTU presidents,

> [W]ould you feel that you could yield anything to [educators'] knowledge and experience, or to their educational theories, as to the best methods for giving [temperance] instruction?
>
> Would you take a half loaf in harmony or demand the whole in possible discord?
>
> Would you make concessions of method where principle was not involved or would you take the position that the method itself is a principle?
>
> Would you consider that, as State President you must adapt the standards of the several National departments to State conditions or would you consider each National Superintendent supreme in her own department? Where . . . is the final Court of Appeal?[16]

Worried that state WCTU officials would not judge her kindly, meanwhile, Mary Hunt appealed to the highest court of all: national president Lillian Stevens. "[E]nemies are boasting that the backward movement . . . is to sweep the country beginning in New England," Hunt pleaded in August 1901. "Dear Mrs. Stevens, God helping us we must prevent this by forewarning our forces." Hunt asked to address the state WCTU convention in Stevens's native Maine, where a recent bill to add STI textbook and penalty provisions had gone down to defeat. Perhaps, Hunt suggested, her presence could help jump-start another drive for a tougher measure—and stave off any movement for détente with state educators. Yet Stevens demurred, barring Hunt from the Maine convention. Stevens had always opposed statutory penalties, which seemed to insult educators' professional acumen. If state WCTU leaders now wished to broker a deal with these schoolmen, she added, they had every right to do so—without interference from Mary Hunt.[17]

The following spring, Katharine Stevenson announced that twenty-seven out of thirty WCTU state presidents in her survey favored "compromise for the sake of harmony" with STI's educational critics. For some respondents, no doubt, this position was simply pragmatic. In Connecticut, for example, WCTU president Cornelia B. Forbes told Mary Hunt that "a fight to the end" would "lose all." Meeting with Wilbur Atwater in Middletown, she did manage to rebut the chemist's charge that STI taught mistruths about alcohol.

"[S]cientific research on the other side is just as strong," she told him, citing Hunt's handful of friendly physiologists. "This is a question where doctors disagree." Yet she was unable to invoke any such authorities from the field of education, where every prominent expert seemed opposed to Scientific Temperance. "You have no idea *how strong* the feeling is," Forbes wrote Hunt, after conferring with state educational leaders. "[I]t is *so* great that it will be *impossible* to save our law unless in *some way* we can come together and yield a point." Shortly thereafter, Forbes joined hands with several schoolmen to endorse a weaker STI measure. It omitted primary schoolers from the subject, in accord with one educator's insistence that "little, happy children" should be shielded from "unnecessary knowledge of evil."[18]

Other WCTU leaders agreed, echoing schoolmen's liberal critique of Scientific Temperance. As several historians have reminded us, romantic notions of child nature took root mainly among educated and well-to-do Americans. Since state presidents stood above the WCTU average in wealth and schooling, they were also more likely to embrace new theories like "culture epochs" and "wholistic" study. "Is there a magic spell about the text-book in physiology which does not inhere in any other text-book?" asked Stevenson, condemning STI's rigid emphasis upon rote learning. "We need to have all sides of this great subject presented in order to produce [the] highest development of the child's entire being, physical, moral, and spiritual." Scientific Temperance threatened this healthy development by harping upon the dangers of drink, Stevenson argued. "Too often we try to lead the children into temperate lives by holding up intemperance only," added another WCTU leader, who doubled as a local school superintendent. "[T]hese lessons can be most effectively taught by placing before the child . . . the virtues we would have him incorporate into his own life."[19]

In many ways, of course, this critique harkened back to the passion and perfectionism of antebellum revivalism—and of Frances Willard. Yet child-study gave these venerable doctrines a new scientific imprimatur, as Willard recognized near the end of her life. Indeed, WCTU leaders often invoked her sacred name in their calls for compromise with enemy schoolmen.[20] "The educators ask this on purely pedagogical grounds," wrote Katharine Stevenson in May 1902. "In no other branch does an outside agency step in and say to [them], 'You shall do this thing in their way.'" Here was a perfect epigram for Willard's style of liberal reform, in which lay citizens set the contours of policy but experts carried it out. To sustain Scientific Temperance, then, Mary Hunt would have to persuade local WCTUs to "step in" and enforce it—even against the objections of their own state and national leaders.[21]

Hunt's Response: Returning to Localism

As in the past, Hunt could not count upon official WCTU publications to circulate her views among local unions. "What fallacies that article of Mrs. Stevenson's contains!" Hunt wrote a friend, after Stevenson called for compromise with enemy educators. "As usual she is not clear as to facts, and the subject is closed for the Union Signal." Once more, then, Hunt distributed her own literature—and state WCTU leaders tried to squelch it. After educators replaced scientists as STI's main critics, for example, Hunt produced a series of leaflets defending its classroom methods. The leaflets also asked local unions to convene special conferences on the question—first for members alone, then for school boards and teachers. Yet few such meetings seem to have occurred, thanks at least in part to Hunt's opponents atop state WCTUs. In Michigan, for example, union president Mary Lathrap ignored the new STI plan altogether. A loyal ally of Frances Willard, she called instead upon local women to visit their respective schools—and post Willard's portrait. Two years later, just nine of sixty-three Michigan WCTUs reported sponsoring Hunt's suggested conferences on Scientific Temperance.[22]

Meanwhile, Hunt kept up a steady stream of personal letters with STI superintendents in the field. Especially after 1900, however, this correspondence emphasized a new theme: local discretion. Earlier, Hunt had stressed the need for every WCTU to follow the same "Schedule and Standard" in enforcing Scientific Temperance. Indeed, she rated each union's "efficiency" based upon its adherence to her guidelines for school visits, textbooks, and so on. As STI came under more concentrated fire from experts, however, she insisted that each WCTU should decide for itself how to use the "ammunition" she sent them. To aid Tennessee women in an appearance before their school text commission, for example, Hunt happily provided circulars that criticized the state's required STI books and recommended alternatives. Yet she refused to tell the workers whether they should suggest a new series or merely condemn the current one. This choice "depends largely upon local conditions with which you, of course, are better acquainted than I," Hunt wrote Tennessee's STI superintendent, "so that you can better judge what would be the effect of combining the two requests."[23]

Similar caveats marked Hunt's counsel about swaying teachers, who increasingly sided with school administrators against STI. "It is difficult for us at this distance to do what you who are on the ground could do by personal talks and visiting the schools," wrote Hunt secretary M. McDuffee, responding to an Oregon woman who complained about teacher resistance in her town. Hunt even declined to advise an Illinois STI worker who inquired about invoking her state's penalty provision against derelict instructors. Such hardball tactics threatened to antagonize teachers even further, Hunt acknowledged. Yet she could not instruct the Illinois woman "without knowing the persons with whom you have to deal," she wrote. "[M]ethods have to be adopted to 'temperament.'"[24]

Finally, even Hunt's letters to state STI lieutenants emphasized the autonomy and authority of local WCTUs. When a new STI superintendent assumed the reins in Washington State, for example, Hunt warned her against trying to effect too much from above. "[S]pecial responsibilities rest upon the local union," Hunt wrote, "as it is with them that the enforcement of the temperance education laws ultimately lies." Taking Hunt's cue, state STI chiefs frequently denied constituents' requests for standardized, step-by-step instructions. "Your State Superintendent cannot give detailed methods for working in every locality, because she views the field only at large, and cannot always meet the difficulties which are sure to arise in various forms," wrote New York's Lytie Davies. "So let every superintendent be on the alert, remembering that 'new occasions teach new duties,' and she must be able to meet them *as* and *when* they come."[25]

Why It Worked: STI Versus Liberal Reform

Given the national prominence of her expert foes, Hunt wondered in private whether untrained, poorly educated WCTU women would rise to the challenge.[26] But they rallied eagerly behind STI, often *because* it required their ongoing vigilance and participation. "We are largely responsible for the law," declared a Kansas STI worker in 1903, "and the promulgation of its principles devolves upon us." Here she contrasted Scientific Temperance with other "positive" as well as "negative" Progressive-era reforms, to borrow historian Paul Boyer's terminology. Positive measures like playgrounds and poor relief required the "continuous management" of experts, who alone possessed the skills and knowledge to execute them. Negative reforms sought victory in a single, dramatic stroke: the classic examples were prohibition and antiprostitution initiatives, which typically aimed to eliminate rather than regulate a "social evil." Despite their differences, moreover, both types of reform rendered lay participation obsolete. Vesting authority in professional managers, positive reform seemed to operate by what Boyer calls "remote control." Negative measures appeared to operate without *any* control, promising immediate—and symbolic—redemption from a host of sins.[27]

By contrast, Scientific Temperance enlisted thousands of laypeople in deliberating, designing, and administering state laws. All human behavior has some symbolic dimension, of course, and STI was no exception: to commemorate legislative victories, for example, Mary Hunt often staged "patch-cutting" exercises where WCTU women would remove a black cloth from their state's map to reveal the pure white underneath. After that, however, Hunt enjoined these same women to enforce the measure within their respective districts. Hence STI engaged them in the nuts-and-bolts spadework of political life, not just the rich symbols that surrounded it. Across America, women seemed to sense the difference. "The W.C.T.U. is a body of policemen," a proud New York STI worker wrote, "self-charged . . . to watch out for the enforcement of this scientific temperance law." Whereas other reforms might require expert intercession, an Iowa WCTU woman added, "[t]his work is our very own."[28]

Meanwhile, STI's decentralized system of authority merely cemented local allegiance to Mary Hunt. Hunt ruled her Massachusetts headquarters with an iron fist, inspiring more fear than love among her personal assistants and secretaries. Indeed, she often forgot "that there were limits to the physical endurance of her workers," as one aide recalled. Given the enormous diversity of WCTUs on the one hand and schools on the other, however, Hunt could not hope to exact the same discipline from STI workers in the field. Less out of choice than necessity, perhaps, she declared her faith in them—and won their devotion, forever. For at the exact point when experts began to dominate American education, Mary Hunt told common citizens across the country that they could still determine what their children learned. Upon her death, an STI superintendent from rural New York wrote:

> There were griefs and many heart aches, there were delays, indifference, and opposition where unexpected, even from dear ones, but there was always Mrs. Hunt to write to when things seemed to come to a dead standstill. . . . Sometimes I almost feared to send her such long detailed ones as I poured out, but she always responded so promptly, giving attention to every point making my way so easy and plain. . . . I believe the name of Mrs. Mary H. Hunt is worthy to be written above every name of woman except that of her who had the nurture and care of the early years of the Lord Jesus.

This woman, it must be emphasized, had never met Mary Hunt. Other correspondents likened her to Wesley, Luther, and Saint Paul. Mary Hunt, meanwhile, seems to have adopted an even higher model. "I come not to bring peace, but the sword," she wrote a friend in August 1902. "We, too, must have our Gethsemanes and Calvaries before reaching ultimate harmony." With a rejuvenated local army behind her, Hunt was ready for her final showdown that fall with Katharine Stevenson, Lillian Stevens, and the rest of her foes atop the Woman's Christian Temperance Union.[29]

A Double Victory

The clash came at the Union's 1902 national convention, held in Stevens's native state of Maine. Already there was talk of other unions following the lead of Connecticut, where WCTU officials had recently brokered a deal with schoolmen to water down their STI law. That summer, moreover, Stevenson released her report claiming that all but three state WCTU leaders supported exactly such a compromise. "The National W.C.T.U. must pronounce upon the principle involved in the Massachusetts controversy," one Hunt associate wrote, referring to Stevenson's détente with Bay State schoolmen, "and I believe it will pronounce correctly." Another ally was less sanguine, noting glumly that "the W.C.T.U. seems to be forsaking Mrs. Hunt." For her own part, Hunt braced for the worst. "So much abuse has been mine of late," she wrote, "that I supposed any *very* little power I once had to win was gone."[30]

But the convention proved her wrong, underscoring the vast gap between state leaders and the local rank and file on Scientific Temperance. When Hunt submitted a resolution "to resist every effort to weaken the laws," nearly every state president rose to condemn it. One critic after the next invoked "states' rights," complaining that Hunt's resolution would prevent them from adapting STI to "local conditions." But local *delegates* rallied to her defense, especially after Connecticut WCTU chief Cornelia Forbes moved to refer the resolution to executive committee. Forbes's motion sought to muzzle the rank and file, since this committee was composed exclusively of national officers and state presidents. But it actually had the inverse effect, galvanizing angry delegates behind Mary Hunt. By an overwhelming majority, they defeated Forbes's motion and endorsed the resolution against compromise with enemy educators. "I should have been helpless, without you, the great army," Hunt told the delegates. "Your support has not been the blind loyalty of following your leader, but of intelligent adoption of what the judgment approved."[31]

In state legislatures, likewise, local STI workers held fast to their original laws—and fought off two successive waves of attack. The first corresponded to Wilbur Atwater's initial publications in 1899, when at least four states considered repeal bills. "We set the white ribbon police force up and down the state to work," wrote an STI superintendent in New York, site of the most serious assault. "[T]he friends of the bill never had courage to press it, and it died in committee, strangled by white ribbon."[32] Two years later, the Connecticut debacle set off six more statehouse challenges. But these attempts to enfeeble STI also came to naught, despite a schoolman's prediction that the new Nutmeg State law "marked the beginning of a reaction in legislation" across the country.[33] The only other state to weaken its STI law was Minnesota, where a revision of the education code on the eve of Hunt's death in 1906 slashed several STI requirements. Yet this defeat was counterbalanced by developments in North Dakota, which had toughened its statute the previous year.[34]

Thus Scientific Temperance emerged essentially unscathed from its turn-of-the-century turmoil, belying both contemporary predictions and retrospective accounts. Most recently, historian Philip J. Pauly has claimed that Hunt's influence over STI eroded during the final years of her life. If anything, however, the battle with WCTU leaders strengthened her hand. In 1905, the national convention resolved to resist "any scheme of education which will take this study from the lower grades, or modify in any way the utmost warning." Just three years earlier, a similar motion had sparked hours of argument; now, it passed with barely a sound. Supporters included Stevens, Stevenson, and a host of other leaders who had long opposed Mary Hunt.[35]

In other ways, however, they continued to make like difficult for her. Repeatedly, national officers refused Hunt's request for a larger annual appropriation from the WCTU. "[A]s all states now have STI laws," assistant recording secretary Elizabeth P. Anderson told her in February 1906, "the expense of the work of the department can not be so great." All that remained was "the work of reserch [sic] and investigation," Anderson added, which "might be carried on with less expense, less clerk hire, etc. etc." To be sure, Hunt spent much of her time researching and writing rebuttals to expert attacks. But she also corresponded directly with local STI workers in the field, which required far more energy and expense. During a single five-day stretch in 1906, for example, Hunt received letters from twenty-nine different states; without funds to hire secretaries, she could never answer all of them. Clearly, she concluded, critics like Anderson "have no conception of what is being done here."

In at least one manner, however, Anderson understood Hunt's efforts all too well. Elsewhere in her letter, she charged that Hunt drew upon a second, clandestine source of money: textbook royalties. "It is claimed that you have an income which has grown out of the work," Anderson wrote, "which you have never been willing to explain." Two months later, upon Mary Hunt's death, the WCTU would receive a full account of this textbook operation. The revelations helped spark the dissolution of Hunt's dense local network— and, with it, of Scientific Temperance Instruction in America.[36]

PART III
REQUIEM: ABANDONING THE FIGHT, 1906-1925

The welfare of society . . . depends upon the welfare of its individual members. The social, domestic, industrial, and economic effects of alcoholic drinks are all a result of the physiological effects.
CORA F. STODDARD, 1908

[E]very new subject which has gone into the schools, has had to be started and pressed from the outside. . . . [I]t seems to be a bit appalling that our own women should fail to avail themselves of this opportunity.
CORA F. STODDARD, 1924

Shortly after Mary Hunt died in 1906, Scientific Temperance would expire as well. Amid revelations that Hunt had received secret textbook royalties, WCTU leaders disbanded her local army of assistants and declared a new era of cooperation with physiologists, educators, and other erstwhile critics of STI. A small circle of loyalists continued Hunt's research efforts at her Massachusetts headquarters, accenting alcohol's effects upon the individual drinker. Without a grass-roots network to transmit it, however, this material rarely reached American classrooms. By the time of America's entry into the First World War, if not earlier, Scientific Temperance had all but disappeared from American schools. In its place arose a broader "hygiene" curriculum, reflecting the growing liberal emphasis upon social relationships and responsibilities over personal character and consequences.

After the war, former Hunt secretary Cora Stoddard tried briefly to revive her mentor's physiological approach. By that time, however, it was too late. Dispensing altogether with issues of individual knowledge and motivation, a new generation of health educators sought to instill "efficient habits" in schoolchildren. Since most WCTU women had abandoned STI years before, meanwhile, they lacked both the understanding and the confidence to reinstate it. During these same years, though, thousands of other lay activists were converging upon schools to shape curricula. Most notorious are the antievolutionists, who won bans on the teaching of Darwin in several state legislatures and in hundreds—perhaps thousands—of school districts. Yet a huge variety of lesser-known groups also charged into schools "from the outside," in Stoddard's phrase, altering everything from history and English to physical education and foreign-language instruction. Hardly an aberration, then, Scientific Temperance pioneered a form of lay influence that would extend across our century. As the largest and most successful movement of its kind, it provides a unique window onto the dilemmas and dangers of popular curricular control in an age of expertise.

7

"LOCAL INFLUENCE" ON THE WANE: SCHOOL HYGIENE AND THE DEATH OF SCIENTIFIC TEMPERANCE

In June 1908, the *Atlantic Monthly* published a lengthy attack on Scientific Temperance Instruction by one of America's leading liberal reformers, William H. Allen. Allen had recently helped found New York's Bureau of Municipal Research, which was dedicated to a wide diffusion of knowledge about the functions and foibles of city government. He laid special emphasis upon problems of public health, hoping that informed voters would press municipal leaders for cleaner streets, water, and air. Yet in schools, Allen maintained, STI prevented the nation's youngest citizens from receiving the information they needed. "Until quite recently the term *school hygiene* stood for one idea, namely, compulsory instruction in . . . the evils of alcohol and nicotine," Allen wrote. But poor diet, physical defects, and "unhygienic living" posed far more immediate threats to America's children, he added. So school hygiene must widen its scope, Allen argued,

> to make of every school child a militant teetotaler who abstains from measles, typhoid, scarlet fever, tuberculosis, dirty streets, and impure air as well as from alcohol and tobacco; to arouse as much indignation against waste of baby-life because of unclean milk or ignorant care as against the pipe and decanter; to inculcate a love of self-control and self-respect that will operate against coffee and tea and gormandizing as well as against cocktails and cigarettes; . . . to teach [that] the violation of natural law by means of corsets, high-heeled shoes, cosmetics, needless visits to physician and drug store, or unnatural living, is antisocial even though the citizen never touches alcohol or tobacco.[1]

Allen added a special jab at the "army of crusaders" in the Woman's Christian Temperance Union, "ready at a moment's notice" to defend STI. Ever since Mary Hunt's death in 1906, however, this female fighting force had been in retreat. Echoing Lillian Stevens and Hunt's other rivals atop the WCTU, new STI superintendent Edith Smith Davis announced that the WCTU would henceforth cooperate—not clash—with American educators. Led by longtime Hunt secretary Cora Stoddard, meanwhile, a small core of loyalists vowed continued vigilance on behalf of Scientific Temperance. Since she lacked Hunt's dense network of local assistants, however, Stoddard would have to rely upon the same methods as Allen's municipal bureau: research, publicity, and advertising. "There are five great leaders of public thought and action: the press, the school, medical opinion, the pulpit, the platform," Stoddard declared in 1908. "To link these leaders in the education of the whole people . . . is the most important work of the present."[2]

The effort failed. Ignoring Stoddard's barrage of pamphlets and posters, school districts across the country gradually reduced STI to a minor sidelight of a much wider subject: school hygiene.[3] Like the nation's burgeoning prohibition movement, hygiene embodied the growing liberal emphasis upon "social bonds" rather than individual character.[4] A child's illness or even intemperance reflected a debased environment—whether a filthy tenement, an unventilated school, or a saloon-infested neighborhood—and could be "cured" only by concerted, collective assaults upon these hazards. Educators worked closely with state and municipal health officials, forging new professional organizations to collect information—and to press for reform.[5] Encompassing nutrition and vaccination as well as sanitation and medical inspection, school hygiene had all but replaced Scientific Temperance by the outbreak of World War One.

As we saw in the Introduction, historians have generally offered two explanations for this rapid professional ascendance at the outset of our century. Following the hygienists themselves, one interpretation would attribute their victory to the new demands of "our highly organized civilization," as Columbia biologist Maurice Bigelow argued. Appropriate for simpler times, perhaps, Scientific Temperance simply could not address the complex health dangers spawned by immigration, industrialization, and urbanization. Another perspective would stress the "careerist ambitions" of school and health officials, who manipulated public fears and pocketbooks to enhance their own standing. In 1904, in fact, a committee of hygienists spelled out their strategy for capturing the curriculum. Loosely worded STI laws "leave ample openings for other text-books, methods, teachings, to any . . . who *have the capacity* to take advantage of such openings," the committee underlined. "[B]y the local influence of hundreds of thousands of members of the Woman's Christian

Temperance Union, exerted on school authorities, the majority of [textbooks] now in use bear its indorsement. If the American Medical Association should indorse other books [and] exert greater influence . . . according to the law of force, it must prevail."[6]

Why, then, did the WCTU's local influence decline? The question shifts our attention from experts to laypeople, whom most scholars still omit from the study of professionalization. New sanitation problems widened the demand for school hygienists, to be sure, while revamped organizational techniques helped enrich their stature. Just as obviously, however, neither explanation can account for the abrupt *timing* of their triumph. Under "the law of force," as they called it, hygienists could never have forged a new curriculum unless laypeople—in this case, the WCTU—forsook the old one.

This outcome was hardly inevitable, as some theories of expert power would imply. Instead, it was touched off by the death of Mary Hunt and a series of chance events that ensued: the revelation of her textbook royalties, the revocation of these payments, and a rift between WCTU leaders and her advisors. These contingencies—not the inexorable drive of professionalism—account for the sudden decline of Scientific Temperance. Most members of our own profession shy away from such claims, as historian Michael McGerr has noted. "Contemporary historiography has little room for chance and contingency," McGerr writes. "[T]hings happen because they are supposed to happen, because of the configuration of social forces."[7] Sometimes, though, things happen by accident. Like the fly eluding a spider, the past often escapes the webs of theory that we spin around it.

Physiology, Hygiene, and Scientific Temperance, 1879-1906

In 1895, a New York teacher journal published a lengthy satire entitled "The Hygiene Examination of the Future." At the outset, a teacher asks "Johnnie" whether he has been vaccinated against smallpox, diphtheria, and cholera. She then makes him promise not to exchange sponges or pencils, both considered common modes of tubercular transmission. A willing pupil, Johnnie also pledges to fumigate his books with sulfur and sprinkle his clothes with "chloride of lime" at least once a week. Finally, the teacher pronounces him fit. "Johnnie, you have met the first requirements of the modern sanitarians," she declares. "[C]limb over yonder rail, occupy an isolated aluminum seat and look at that awful picture on the wall of a drunkard's stomach till I get time to show you about making P's and Q's as your next lesson."[8]

As the parable indicates, critics often linked Scientific Temperance to school hygiene. Both subjects had roots in a distinctive tradition of "Christian Physiology," dating back to the massive waves of reform that swept the antebellum North. According to health spokesmen like Sylvester Graham, whose famous cracker still bears his name, a knowledge of digestion and other bodily functions would help ward off sins of the flesh. In the nation's new common schools, likewise, Horace Mann demanded "a study of our physical frame" as a way to demonstrate "the laws and conditions of Health and Life."[9] Building squarely upon this idiom, fellow New Englander Mary Hunt framed her STI laws to require "physiology and hygiene, with special reference to alcohol and other narcotics." In lessons on drink and tobacco as well as diet and tight lacing, Hunt argued, the best route to a sound body was "the impartation . . . of scientific facts" to the mind. For the most part, doctors and scientists agreed. In 1882 and again in 1886, for example, the American Medical Association endorsed mandatory STI "in connection with . . . physiology and hygiene."[10]

By the 1890s, however, new developments in medical science had changed the meaning of these disciplines—and placed STI on the periphery of both. As we saw in Chapter Three, a fresh laboratory ethos led physiologists to question Hunt's bedrock claim that alcohol was poison. Meanwhile, dramatic discoveries of harmful bacteria boosted the scope as well as the authority of American hygiene. Unlike their antebellum forebears, who cited Scripture as much as science, a new generation of "modern sanitarians" based their claims squarely upon microbes. Vicious but invisible, these germs could be anywhere—on clothes, on books, even on pencils. In school, then, the sanitarian's most pressing job was to prevent their transmission. Across the country, medical societies and state health boards forged new committees to lobby school leaders—and to fight for new instruction about communicable disease. "[P]ublic schools are just now . . . in the process of readjustment between classic aims and twentieth century needs," declared one such panel in 1903. "It is in such a transition period, before new features are 'settled,' that favorable opportunity for bettering instruction in hygiene offers itself."[11]

Turning their gaze to American classrooms, however, sanitarians found a huge obstacle in their path: Scientific Temperance. According to Mary Hunt's "Standard of Enforcement," STI required just four lessons per week for ten weeks; at least in theory, the rest of the school year was available for vaccination, medical inspection, and warnings about contagion. Likewise, Hunt asked that a quarter of the pages in her endorsed primary school textbooks—and twenty pages of her high-school texts—address alcohol; again in theory, there was plenty of room for other topics. In practice, however, STI tended to crowd them out. "[T]he teaching of physiology and hygiene, which ought to be of great help in the everyday life of the pupil, is becoming synonymous with teaching the effects of alcohol and narcotics," one New York committee reported in 1902. "[I]n some schools, the text-books in the subject are called by the pupils 'the liquor books.'"[12]

Even worse, sanitarians complained, such instruction as did occur stressed physiology over hygiene—that is, the reasons for healthful living over the rules that governed it. "It is far more important to us to know how to keep the eye, or the stomach, or the heart . . . in health, than it is to know the structure of the organ, or to know how it carries on its processes," declared George Groff, president of Pennsylvania's state health board. In a 1903 address, revealingly entitled "Physiology *versus* Hygiene," Groff went so far as to argue that the latter could be taught without any reference to the former. Children should learn proper health habits "incidentally," he added, as the "psychological moment" arose: when it rained, for example, their lesson would stress the need for dry clothing. Only in high school, however, should they study the physiological rationale for it. "The teaching of hygiene [is] to a certain extent a farce," added biologist Benjamin Lee, chair of an American Public Health Association panel on the subject. "[T]he immature mind of the child should not be overburdened by memorizing descriptions of bones and muscles."[13]

Here sanitarians echoed the romantic theories of child-study educators, who also condemned Scientific Temperance for harping upon sin—and for hindering the natural development of the psyche. Indeed, much of the hygienic critique—especially the emphasis upon students' "psychological moment"—was a gloss on the old Froebelian formula that interest should determine subject. Just as liberal schoolmen joined hands with like-minded theologians and physiologists to fight STI, then, so did they seize upon new trends in school hygiene. "We should teach [children] the laws of health, including many things which the present obnoxious law does not require," wrote New York superintendent Charles R. Skinner, during his unsuccessful drive to dilute his state's STI measure. "More of our children become incapacitated through the effects of illy ventilated school-rooms than from drinking beer or whisky." Journals like Nicholas Murray Butler's *Educational Review* opened their pages to the sanitarians, whose extensive surveys documented the "perfunctory" and "superficial" nature of school health instruction—and the need for university departments in the subject. "From such fountain-heads," one hygienist rhapsodized, "teachers, health officers, sanitation engineers, physicians, and investigators would take their inspiration."[14]

Out in the schools, however, WCTU women took their inspiration from Mary Hunt—and blocked this attempt to bury Scientific Temperance in a thick haze of hygiene. The sanitarian challenge followed the same liberal reform approach that historian Robyn Muncy has outlined: experts investigated a problem, publicized the results, lobbied for a solution, and demanded their own appointment to administer it.[15] Hunt, on the other hand, insisted that laypeople could—indeed, must—interpret and monitor expert prescriptions. Following her own model of reform, she cited other professionals to counter hygienists' critique. Then she channeled their claims through a dense grass-roots network of assistants, whose personal intercession preserved STI.[16]

Hunt's first step, as always, was to enlist friendly experts. When the American Academy of Medicine appointed a panel to study school hygiene, she dispatched aide Cora Stoddard to contact its members. Stoddard also wrote an urgent letter to Northwestern physiologist Winfield Hall, an AAM member who had helped stave off earlier attacks upon STI's scientific claims. "Please watch the newspapers," Stoddard urged Hall, "and correct anything that needs explanation or correction." For her agents in the field, meanwhile, Hunt prepared literature to rebut the new sanitarian foe. After a teachers' association called for curricula "based upon hygiene rather than physiology," for example, Hunt's office helped design and distribute a point-by-point reply. "Hygiene with no physiology [is] mere exhortation," the pamphlet declared. "It would be as reasonable to propose that this age of steam and electricity should go back to the stage coach . . . as that moral suasion should now take the place of physiological temperance instruction in our public schools." Here, as in her battles against scientists and educators, Hunt turned the experts' arguments against them. When sanitarians labeled STI a relic of simpler times, she affixed the same charge to their own emphasis upon "ethical" rather than "scientific" training.[17]

Simultaneously, of course, she also stressed STI's majoritarian roots. "It was not the teachers as a class, but the people, the source of power in this country, who demanded [STI]," the pamphlet concluded. "The laws that these teachers would practically nullify were enacted in response to the petition of millions and millions of fathers and mothers." Out in the schools, too, her assistants oscillated easily between expert and popular appeals. Responding to an educator's claim that STI's scare tactics encouraged drink, an Indiana STI worker pointed out that other hygienic instruction—which the educator supported—employed the same strategy. Surely, Ida F. Dickerson argued, expert schoolmen would not discontinue lessons about the "evil effects" of impure water and poorly ventilated rooms, "for fear that the little Tom Sawyers will straightaway do the very things that parents and teachers have advised them not to do." Women like Dickerson were the sole bulwark blocking these other subjects from swallowing up STI, as one Massachusetts schoolman admitted. "If we were left free from outside pressure, we could develop a much better course," he wrote. "We should not teach [STI] to small children." Only when Mary Hunt passed away would this "outside pressure" dry up, allowing school hygiene to drown out Scientific Temperance.[18]

Coming Apart I: The Smashup of 1906

On 24 April 1906, following a lengthy illness, Mary Hanchett Hunt died. A few days later, her closest allies gathered in Boston to discuss the continuation of her work. Included were several members of Hunt's advisory board, who favored abandoning the WCTU and forging a new organization. Just before her death, minister E. O. Taylor reported, Hunt had told him "she felt handicapped in the WCTU" and "found it necessary to go outside for influence." The Union, he reminded listeners, was "no authority to quote." So Taylor suggested a separate scientific institute, which would collect research by the best experts on the subject. "Physicians call everywhere for authoritative information and need such an organization," added inebriest and STI stalwart T. D. Crothers. "Such a one is coming whether we form it or not."[19]

By enlisting experts like Crothers, Hunt had likewise sought to establish an "authoritative" pedigree for Scientific Temperance. But she funneled their findings through thousands of WCTU assistants, leavening expert authority with a popular yeast. "Mrs. Hunt had the philanthropist's interest in . . . teaching children," argued former Hunt secretary Emma Transeau, who had answered letters and distributed literature to STI workers in the field. "Scientists would perhaps not see the importance of this." Turning a deaf ear to Transeau's calls for grass-roots activity, Hunt's advisors proceeded to endorse a strictly research-based organization. The WCTU would be invited to join their new "Scientific Temperance Federation" (STF), receiving access to Hunt's vast array of temperance materials. Yet the STF would not mobilize local WCTU women to disseminate this information. With the WCTU's influence "at a standstill," Taylor maintained, any connection to the Union could only hamper the sleek, "professional" image that his new enterprise strove to project.[20]

Of course, this plan presumed that the STF would own or control Hunt's temperance collection. After her death, however, national WCTU leaders claimed otherwise. Since Hunt herself had credited "the machinery of . . . local unions" for her success, one official argued, any STI materials she amassed should revert to the mother Union. Hoping to head off such claims, STF members prevailed upon Roy Hunt—Mary's grandson and only heir— to bequeath her books, file cabinets, and typewriters to a quasi-secret corporation called the Scientific Temperance Association. Yet this move merely heightened the controversy surrounding her estate, as Hunt's advisors later acknowledged. For in transferring these materials to the STA, they were forced to admit what had long been rumored: Mary Hunt earned royalties on the textbooks she endorsed. The Scientific Temperance Association had been formed in 1886 to collect the moneys and distribute them to Hunt's Hyde Park headquarters, where they helped defray the enormous costs of her STI operation.[21]

News of this arrangement sent shock waves across the WCTU in the summer of 1906. As a flurry of "trust" exposés heightened American concerns about business corruption, one WCTU wag compared the STA to Standard Oil. Others reacted like jilted lovers, stunned that their "dear Mrs. Hunt" could have deceived them. Such feelings were fueled by Hunt's enemies atop the Union, who feigned shock at the revelations. In fact, they had known all along. Hunt told Frances Willard about the STA as early as 1892, repeating the description in a 1903 letter to Lillian Stevens. To augment the STA's aura of mystery and malfeasance, however, Stevens now insisted that "the National WCTU did not know of the existence of the association" until Mary Hunt died. The entire affair cast Hunt in the worst possible light, seeming to confirm her foes' longtime charges of greed, vanity, and dishonesty. It also tarnished her former advisors in the Scientific Temperance Federation, many of whom sat on the STA as well. To protect their new research organization—not to mention their good names—these men now set out to explain the origins and operation of Hunt's textbook kingdom.[22]

According to several statements they released in 1906, the Scientific Temperance Association was born during a Missouri book dispute two decades earlier. Before that, Hunt had rejected renumeration for the editing services she provided to A. S. Barnes and other friendly publishers. In 1886, however, Barnes quietly pared the sternest antialcohol passages from Hunt's endorsed Pathfinder #1 book and released it as a special "St. Louis edition." To protect the total-abstinence tenets of such texts, Hunt realized, she would have to assume "author's or editor's rights of control"; yet to retain her own veneer of womanly virtue, she would also have to avoid any "imputation of self-interest." Hence she obtained legal claims to her endorsed material and transferred them to an independent board of trustees, the Scientific Temperance Association.[23]

These claims, Hunt's advisors added innocently, included her "right to compensation." Even her enemies acknowledged that Hunt reviewed book proofs with almost maniacal care, refusing to affix her approval until every specification had been met. Indeed, this editorial work proved such a burden that Hunt suffered a brief "breakdown of strength" and entered a New York sanitarium in 1891. Hardly boodle or bribery, then, her book royalties represented payment for services rendered. Moreover, every penny went to subsidize her costly regimen of research, correspondence, publishing, and travel. If it wished, Hunt's advisors wrote, the WCTU could join the Scientific Temperance Association and become a trustee of the materials at her headquarters. It could not assume sole title to these materials, which Mary Hunt had earned by the sweat of her brow.[24]

Yet to Hunt's opponents in the WCTU leadership, any shared ownership was unacceptable. "[W]e cannot put our organization into the hands of men, to say that we must do this and so," New York's Mary Allen told its 1906 national convention. "We do not want that." Overwhelmingly, the Union's executive committee voted to reject invitations to join the Scientific Temperance Federation and the Scientific Temperance Association. Led by Massachusetts's Maria Gordon, a few dissenting delegates argued that the WCTU needed access to Hunt's extensive library and support from her expert allies. "This resolution cuts us free from those who are our best friends," Gordon pleaded. "I think it is a suicidal act." Nonsense, replied Allen, noting that most of Hunt's materials were available elsewhere. "The Woman's Christian Temperance Union can create its own bureau," another delegate declared. "We are perfectly competent to do it."[25]

Coming Apart II: Edith Davis and the "New" STI

Like Hunt's advisors, indeed, many WCTU leaders now viewed research as the only remaining task for STI. They proceeded to refashion the union's Department of Scientific Temperance Instruction into a clearinghouse for alcohol information, resembling nothing so much as the Scientific Temperance Federation in Boston. At the helm of the department sat a new superintendent, Edith Smith Davis, whose selection reflected an emphasis upon "research and investigation" rather than local organization. This "Leader in Scientific Temperance Instruction" (as the *Union Signal* called her) served only briefly as STI superintendent in her native Wisconsin. Yet as a writer "of marked ability" and a speaker of "several languages," the *Signal* added, she was "in every way competent" to assume the national post. She had even taken a postgraduate course at Wellesley with Alice Freeman Palmer, a close friend of Frances Willard and a longtime critic of Scientific Temperance. Clearly, academic grooming—not grass-roots legwork—constituted Davis's chief credential for the job.[26]

Atop the WCTU, officials called upon local unions to lend Davis "the same loyal and effective help" they had given to Mary Hunt. Yet neither Davis nor her superiors envisioned a continuous, participatory role for these women in monitoring or administering Scientific Temperance. As we saw in Chapter Six, WCTU leaders like Lillian Stevens had long pressed for a compromise with STI's schoolman critics. The death of Mary Hunt gave them a golden opportunity to achieve it, provided that her local army of assistants stayed clear. So Stevens sent all WCTU members a detailed circular about Hunt's textbook operation, hoping to dissuade them from pressing these books upon the schools. In her own report to local unions, meanwhile, Davis instructed them to abandon the zealous enforcement efforts that Hunt had devised. Denouncing "the old idea" of "strict conformity" to state STI measures, with their tedious textbook and time clauses, Davis called for renewed attention to "the spirit of the law" and renewed cooperation with school leaders. These experts—not WCTU women—best understood the proper frequency and method for teaching Scientific Temperance, she maintained.[27]

To underline the shift, Davis announced the appointment of a new "Board of Counselors" that would recommend textbooks and otherwise advise her on scientific and educational matters. Its best-known member was Stanford president David Starr Jordan, perhaps the harshest critic of Mary Hunt and STI during the previous decade. Now that Davis was in charge, however, he sensed that a "commendable change" had come over the WCTU. Jordan especially praised Davis's efforts to influence normal schools and universities, where alcohol-related instruction remained scant. As we saw in Chapter Five, these institutions rebuffed Mary Hunt so routinely that she tried to establish a separate "College of STI" for training teachers. Now that Davis had made peace with educational leaders, however, she could reasonably expect a more receptive hearing. Already, she boasted in 1907, state universities in California and Wisconsin had heeded her "urgent plea" for greater instruction in "physiology and hygiene." By ceding local enforcement to experts, Davis believed, STI would finally gain a place in the academic firmament.[28]

Coming Apart III: Cora Stoddard and the STF

Meanwhile, in Boston, Hunt's advisors were forging a parallel organiza-tion—the Scientific Temperance Federation—to defend her legacy. In 1907, they appointed her erstwhile assistant Cora Stoddard as the Federation's first secretary. Repeating their pronouncements of the previous year, STF leaders emphasized that "stimulation of original research" remained their primary goal. At Stoddard's insistence, however, the STF's Plan of Work also called upon members to visit schools, secure "a definite place in the course of study" for STI, and promote "friendly intercourse" with teachers. The Fed-eration even assumed publication of Hunt's teacher periodical, the *School Physiology Journal*. "There are many women in the WCTU who loved Mrs. Hunt and her work," Stoddard wrote hopefully, "and want to feel that it is going on much as she left it." By enlisting them as "associate members," she added, the STF would disseminate its findings through local school dis-tricts—and preserve the most distinctive achievement of Mary Hunt.[29]

At first, the goal seemed attainable. Even as WCTU officials circulated news of Hunt's book deals, STI workers sent Stoddard expressions of sup-port. "Women are writing me that they 'can not see that our dear Mrs. Hunt did any wrong' and that they must continue to have the [School Physiology] Journal," Stoddard reported in January 1907. "So while we lost some we gain some." As the year wore on, however, new disclosures put her on the defensive. In April, Stoddard learned that publishers had provided royalties on no fewer than twenty-two books—not on fifteen, as was previously an-nounced. Even worse, at least one company had paid its annual sum to Hunt herself instead of to the Scientific Temperance Association. Stunned by these revelations, Stoddard decided that any response would "only stir up more unkind remarks" about her mentor. So she kept quiet, confident that "thou-sands of women . . . still love Mrs. Hunt"—and would soon join the STF.[30]

Out in the field, however, the tide had turned. One local Florida WCTU refused to meet with STF board member E. O. Taylor, simply because he had once been an advisor to Hunt. In Ohio, meanwhile, even Federation sympa-thizers were afraid to voice their feelings. When the STF asked WCTU of-ficial Mrs. J. Frank Smith to write an article for one of its publications, Smith hesitated. "As you know, my heart is still in 23 Trull Street," Smith wrote, referring to the Hyde Park address of Hunt's headquarters, and now of the STF. "Our W. C. T. U. Nationals would no doubt condemn me severely. For the love of Mrs. H. I will try however to do my best." Stoddard hesitated, too, fully aware of the widespread WCTU hatred towards Hunt, the STF, and herself. "Are we justified in asking her to do this," Stoddard scrawled in the margins of Smith's 1909 letter, "when it will cost her animosity and may cost the work in Ohio as well?"[31]

By this time, Stoddard had decided that the answer was "no." She did continue to correspond with WCTU women in Canada, who seemed unaware of the rifts that racked the American movement.[32] Otherwise, though, she abandoned local organization and devoted her energies to a more diffuse task: influencing public opinion. Renaming Hunt's *School Physiology Journal* as the *Scientific Temperance Journal,* she announced that the STF would henceforth function as an "expert bureau"—and would cast off the amateurish aura of the Woman's Christian Temperance Union. "Would not the public be more ready to accept the facts from a separate organization known to specialize in the subject?" she wrote a friend in 1909, rejecting his call for a truce between the Federation and the WCTU. "One can not very well put a half-grown child back into its baby clothes and expect them to fit."[33]

Boasting her own expert allies, meanwhile, Edith Smith Davis sought to avoid any "embarrassing affiliations" with Mary Hunt and her heirs at the STF. At the International Congress on Alcoholism in London later that year, Davis said Stoddard was "so closely associated with Mrs. Hunt's work" that her judgment could not be trusted. Both women ultimately addressed the Congress, each taking credit for the "progress" of "physiology and hygiene" back home. In schools and universities alike, however, this course increasingly focused upon medical inspection, vaccination, and diet—not alcohol. As physiology and hygiene expanded, Scientific Temperance was melting away.[34]

School Hygiene I: Replacing STI, 1906–1913

For almost two decades, of course, doctors and schoolmen had struggled to replace STI with a wider health curriculum. When Mary Hunt died and her local network dissipated, then, they jumped at the chance to complete the coup. Scientific Temperance "was too narrow, limiting as it did to the effects of alcohol and narcotics," wrote Massachusetts education secretary George H. Martin, a longtime Hunt opponent. "The opportunity now exists . . . for a new propaganda in favor of health instruction." In May 1907, Martin traveled to Washington for the first congress of the American School Hygiene Association (ASHA). Accompanying him were Nicholas Murray Butler, Edward Thorndike, William H. Welch, and a host of other medical and educational luminaries, all dedicated to establishing "a broader motive for school hygiene." Within the next decade, literally dozens of other organizations joined the bandwagon. From the American Child Hygiene Association and the Child Health Organization of America to the National Child Labor Committee and the National Child Welfare Association, new and old groups alike fought to expand the focus of health curricula in American schools.[35]

Like Mary Hunt, they began their effort in state legislatures. From 1903 to 1913, twelve states mandated medical inspection for all of their public school students; between 1908 and 1910 alone, twenty-eight passed laws calling for instruction about tuberculosis. Other new statutes required gymnastics, nature study, and dietetics. The language of these measures paid indirect tribute to Hunt, often mimicking her STI laws word for word. In North Dakota, for example, health organizations won ordinances requiring physical education "as a regular branch to all pupils" and tuberculosis instruction "to all pupils above the third year." Both dicta exactly echoed the state's STI measure, as did the added demand that students receive a minimum of four lessons per week for ten weeks in disease prevention.[36]

As always, it is difficult to determine how much of this material actually reached students. Textbooks devoted a greater fraction of their pages to sanitation and other health topics, a reasonable indicator of their rising prominence in the classroom. At least half of American normal schools prescribed coursework in hygiene by 1910, as did forty-seven colleges and universities. For the first time, meanwhile, state and city school systems hired supervisors who specialized in the subject. Like other Progressive-era reformers, these experts jumped easily from the voluntary organizations that demanded new measures to the bureaucracies that administered them. In Maryland, for example, the leader of the Playground Athletic League served as state supervisor of physical education. After persuading the U.S. Office of Education to create a school hygiene division, likewise, ASHA cofounder Fletcher B. Dresslar became its first director.[37]

To be sure, teachers often objected to the new health requirements as loudly as they had denounced STI. "Why should I take a course in school hygiene?" one asked. "It's just something about measuring desks, isn't it?"[38] But others praised the fresh emphasis upon "the care of the body" over "detailed anatomical study," as one Philadelphia instructor noted. In the South, historian William Link has found, hundreds of rural school districts replaced "the study of physiology" with "the study of health." State education departments helped spur the shift, downplaying drink in favor of disease. After a 1911 Mississippi law required classrooms to display "scientific facts" about alcohol, for instance, schoolmen prepared a poster listing three items about liquor—and seventeen about the transmission of tuberculosis.

They even emblazoned it with a Christmas Seal, international symbol of the antituberculosis struggle.[39]

School Hygiene II: From Altruism to Habit

As the content of school hygiene changed, meanwhile, so did its underlying philosophy and rationale. "While physiology may be individual and self-centered, hygiene should be social and altruistic in its point of view," explained William H. Allen in 1908. "The vast majority of us will avoid or stop using anything that makes us offensive to those with whom we are most intimately associated." Insofar as schools still addressed the problem of alcohol, then, they should focus upon its social costs—in poverty, violence, and disease— rather than its physiological consequences. Likewise, students should be warned that dirty hands and poor ventilation put their neighbors, not just themselves, in peril. "Too much emphasis . . . is directed to the personal health of the pupil," wrote Teachers College professor Thomas D. Wood, director of its graduate program in school hygiene, "[and] too little to the health and well-being of members of the home and of the community."[40]

Here sanitarians echoed the mounting liberal accent upon "social bonds" over individual character, as historian Daniel Rodgers has described it. Following G. Stanley Hall and other romantic theorists, critics once worried that STI would stunt the natural development of the child. Now they feared its broader effect upon society, arguing that STI's physiological approach promoted "unwise introspection" rather than concern for the whole. At several New York high schools, for example, student "sanitary squads" set standards of hygiene, performed daily inspections, and meted out punishments to those judged unclean. "In place of the former selfish individualistic attitude . . . [the] student is beginning to think of the other fellow," wrote one health educator, in praise of this peer-monitored system. "[H]e is learning the first lesson of the future citizen—that of cooperation with authority for the common good."[41]

To other hygienists, however, students' motives and attitudes were moot. All that mattered were their habits, which schools could manipulate through games, songs, and other basic techniques of behaviorism. "[O]ur main purpose . . . is to cause the child, and later the adult, to live right habits, to rebel instinctively against offensive and unhealthful practices," sanitarian Louis Nussbaum told a 1912 conference, dismissing another panelist's calls for "altruistic" rather than "egoistic" appeals. If his two-year-old daughter routinely brushed her teeth, Nussbaum added, why should he care about her reasons for doing so? This debate mirrored the larger split between "meliorists" and "efficiency experts" in early twentieth-century educational thought. As historian Herbert Kliebard has pointed out, both groups viewed education as "a self-consciously social enterprise." Yet meliorists retained a vestige of the older emphasis upon character, hoping that schools might temper Americans' avaricious tendencies with a new, collective spirit. Efficiency experts focused more narrowly upon skills and activities: describing schools as factories, they sought to fashion "raw material" into functional "products" via careful measurement of student talents on the one hand and social tasks on the other.[42]

During these same years, medical inspection provided a wealth of data about students' poor diet, posture, and personal hygiene. For efficiency-oriented sanitarians, these statistics posed a problem not of social conscience but of "biological engineering." With such a flawed "input," they asked, how might schools manufacture a better "output"? They seized upon youth crusades, an antebellum tradition that hygienists tried to update for the modern era. "In the past you were a member of a juvenile temperance society where the boys and girls sang songs about the evils of intemperance," wrote University of Pennsylvania sanitarian A. Duncan Yocum in 1910. "You received a lot of trifling impressions, and in the end you had a habit." More recently, antituberculosis societies had enlisted children as "Open Air Crusaders" in the battle royal against consumption. Now sanitarians called for new "militia of hygiene" that would drill the entire panoply of proper health habits, as Clark University's William H. Burnham wrote in 1912.[43]

A protégé of G. Stanley Hall, Burnham would see his dream realized with the 1915 formation of the Modern Health Crusade. More than four million new "pages," "knights," and "squires" pledged their undying loyalty to eleven different "health chores," ranging from handwashing to handkerchief use. Yet the Crusade dispensed with Hall's emphasis upon "subjective conditions"— that is, "the mental attitudes that are dependent upon the individual himself," as Burnham later noted. It also ignored STI, the centerpiece of health education less than a decade before. "It is difficult to formulate a constructive program for the successful teaching of the effects of alcohol," sanitarian J. Mace Andress admitted, "because . . . it is practically impossible to give any kind of training in action." To teach oral hygiene, for example, schools can cultivate "the toothbrushing habit," Andress noted. Yet it would hardly do for them to instruct students in "the drinking habit," since the entire object was to *prevent* its development. For different reasons, then, both meliorist and efficiency-oriented hygienists downplayed STI. One thought it made children focus inordinately on their own selves, while the other could not find a way to practice or apply it as a group.[44]

Responding to School Hygiene I: Davis and the WCTU

These changes seem to have caught Edith Smith Davis and the WCTU by surprise. When Pennsylvania's Yocum introduced a new child hygiene course, for example, the *Union Signal* gushed that "one of our great eastern educational institutions" was teaching "temperance physiology." Yet the course touched only briefly upon alcohol, which it described in terms of "public" costs—in poverty, accidents, and so on—rather than physiological effects.[45] The same emphasis marked the nation's growing prohibition movement, as a host of historians have reminded us. Hardly the pinched preserve of rural conservatives, twentieth-century prohibition echoed the essential liberal theme of individual sacrifice for the common good. "If the use of alcohol were a personal matter, its effects beginning and ending with the individual, the plea of personal liberty might be effective . . . ," explained Stanford's David Starr Jordan. "But the sale of alcohol has its public relations." Specifically, he added, drink promoted vice, filth, and corruption; so it must be removed, regardless of the precise nature of alcohol or the "rights" of its acolytes.[46]

Indeed, Jordan told the National Education Association in 1911, strictly physiological arguments seemed to play into the hands of wet publicists who vilified prohibition as a violation of their freedom. His position placed Edith Davis in a difficult spot. On the one hand, she wanted every student to avoid alcohol for "his own good"—that is, because of its danger to the stomach, liver, brain, and blood. Yet she could not afford to alienate prized academic allies like Jordan and Massachusetts school chief George Martin, who stressed its effect upon society at large. So she tried to strike a compromise between these dual perspectives. In 1910, she began publishing a new journal, the *Temperance Educational Quarterly,* that would present materials from a broad range of viewpoints—not just "scientific" ones. So did Davis's handbook for teachers, *A Compendium of Temperance Truth,* which included chapters about alcohol's effects upon crime, business, and athletics. Using her connections with leading educators, finally, she also won appointment as a "hygiene lecturer" at several colleges and normal schools. Here, too, Davis struck a balance between "social" and scientific approaches as well as between alcohol and other health subjects.[47]

But little of her material actually entered the public school classroom, as Edith Davis came to realize. In deference to educators, she had scaled back the WCTU's local enforcement efforts upon assuming her post. Now she sought a mechanism that would reignite local STI interest without reviving the strict surveillance that schoolmen detested. She seized upon nationwide essay competitions, offering prizes in 1909 for the best STI papers by grade school pupils, teachers, and college students. For a brief moment, these WCTU prize essays did seem to spark more enthusiasm for the subject. Once the novelty had worn off, however, they simply highlighted STI's precipitous decline in American schools. Most districts ignored the essay competition; a small handful deluged the WCTU with requests for literature, indicating that STI was no longer taught there on a regular basis. "[O]ur supreme need is a revival of popular interest in securing the instruction," wrote Delaware's STI superintendent in 1912, in the wake of a failed essay campaign. "I am finding town after town where little or no attention is given . . . and the people themselves do not know what is the state of affairs."[48]

The following year, Davis called upon WCTUs to sponsor Sir Galahad Clubs in every public school. Like the knights of old, Sir Galahad's student legions would pledge to "speak pure, true words" and "do pure, true deeds." Most of all, they would promise to keep themselves clean "both in body and spirit"—and to abstain from the latter, at least in its distilled and fermented forms. By this time, however, many local unions had dropped their STI departments: in Ohio, for example, only forty-seven of eighty-eight counties reported any Scientific Temperance work at all. Even as Davis's clubs aimed to spark new interest in STI, then, they foundered for lack of a grass-roots network to initiate them. So Davis shifted her energy to promoting an annual Temperance Day, which embodied the same type of activity on a much smaller scale. Once a year, usually on Frances Willard's birthday, students would participate in parades, pledge signings, and skits against alcohol. Thanks largely to Davis, ten state legislatures mandated some type of Temperance Day observance by 1918.[49]

Like the essay contests that preceded it, however, Temperance Day signaled the dissolution rather than the rejuvenation of STI. Parades and pledges invoked the venerable American tradition of Gospel Temperance, of course, dating all the way back to antebellum cold-water rallies and most recently revived by school hygienists' Modern Health Crusade. Yet just as the Health Crusade ignored the reasons for proper health habits, Temperance Day activities tended to dissolve all instruction about alcohol in a noisy haze of song and horn. Insofar as they targeted the saloon rather than drink per se, moreover, such events helped focus attention upon prohibition—and away from

Scientific Temperance. Students in Tennessee dressed in white and black paper caps on Temperance Day, bearing the names of dry and wet states for 1918; in Florida, they sang a prohibition rally song to the tune of "Onward, Christian Soldiers." But neither mentioned the physiological effects of alcohol. Embracing hygienists' social arguments as well as their drum-and-trumpet techniques, the WCTU had all but relinquished the subject that started it all.[50]

Responding to School Hygiene II: Stoddard and the STF

From the beginning, by contrast, Cora Stoddard and the Scientific Temperance Federation resolved to fight the hygienic challenge. In a May 1908 column entitled, "Shall Temperance Be Social?" Stoddard gave a firm answer: no. "[T]he welfare of society . . . depends upon the welfare of its individual members," she declared. "The social, domestic, industrial, and economic effects of alcoholic drinks are all a result of the physiological effects." Sensitive students might listen to "public" arguments, Stoddard allowed, but the vast majority of youngsters needed the "personal application" that science alone could provide. As hygienists began to stress habit over knowledge, likewise, Stoddard mounted a vocal defense of the latter. "No candid mind will deny that knowledge does not always prevent evil," she wrote, "but the human mind is so constituted that it is open to conviction by fact, and in the main to act upon it, else why have any education at all?" Indeed, the very gains that sanitarians celebrated—in diet, exercise, and so on—indicated "how powerful a means knowledge is for the correction of hygienic habits," she added.[51]

Here Stoddard borrowed a page from her mentor, Mary Hunt, who had often turned opponents' arguments against them. Yet Hunt forwarded this "ammunition" to assistants in the field, where the real struggle for STI was waged—classroom by classroom, school by school. Now her grass-roots army was gone, leaving Stoddard with plenty of bullets but no guns. "[T]he teaching [of STI] will not be done without local pressure," she wrote in 1911, "and that pressure is not being exerted." To influence schools, then, Stoddard would have to work through the indirect medium of public opinion. "[T]he more general we can make knowledge of the scientific facts so that the school leaders see it as an established and increasingly popular subject," she explained, "the sooner we shall see the schools cheerfully falling into line." Everywhere Americans looked—in streetcars, storefronts, and even stadiums—they should see "publicity" against drink, Stoddard argued. Then schoolmen would see it, too, and Scientific Temperance would be saved.[52]

SCHOOL HYGIENE AND THE DEATH OF SCIENTIFIC TEMPERANCE   135

During the next four years, the STF produced and distributed millions of posters, charts, billboards, and pamphlets. Increasingly, however, these messages stressed the social rather than scientific effects of alcohol. Crime, poverty, and prostitution were much easier to capture on a billboard than muscle fatigue, mental operations, or heart failure. They also resonated more deeply for "the general public," Stoddard noted, since Americans could "see" the dangers of alcohol in their homes and lives—but not in their hearts and livers. "[W]e are between the Scylla of stern science that demands all explanatory and qualifying details," Stoddard later wrote, "and the Charybdis of the great uninformed public who can first understand new truths only in broad lines." More and more, then, she transmitted these facts without any mention of "science" at all. A popular example depicted two miniature houses, one "where the money is being spent for drink" and another "where the same workman spent his wages for the home." Viewers would not need physiology to grasp the obvious point: abstinence from drink assured freedom from want, while alcohol led inexorably to poverty.[53]

Yet as Stoddard often noted, with more than a hint of bitterness, poverty also afflicted the Scientific Temperance Federation. Most other temperance organizations declined the STF's offer of membership, concluding that they could collect their own information for a lesser fee. Nor could the Federation count on book royalties, which the Scientific Temperance Association dutifully collected and funneled to the STF. Already in decline before Hunt's death, purchases of her endorsed textbooks fell off rapidly in the years that followed. Even worse, at least one publisher now insisted that WCTU claims on Hunt's property invalidated all prior royalty agreements with her. Hence the handful of endorsed titles that were still popular would not necessarily generate revenue for the STF, as a worried Stoddard explained. Finally, philanthropists rejected her pleas to assist the STF in its "great popular propaganda" against drink. "Your organization has not the large organized constituency through which such work can be done," one wealthy donor told her. "We would give more largely if you had such a constituency."[54]

Privately, Stoddard admitted that she had erred "in not increasing local machinery"—that is, in failing to cultivate a grass-roots network of assistants. Equal fault lay with Edith Davis and the WCTU, "which in the past was diligent in exercising the local pressure" but "lost its influence" in schools, Stoddard added. Relations remained icy with the Union as late as 1913, when Davis took to an Ohio podium to slam Mary Hunt and her heirs at the STF. So Stoddard turned to the Anti-Saloon League, America's most powerful temperance organization, which agreed to bankroll the Federation in June of

1913. "[I]n the near future the liquor people are going to begin to work the scientific end," Stoddard wrote. "If there is any agency by which we can reach the public . . . with scientific facts, I think we ought to use it."[55]

During the next two years, however, this alliance merely accelerated the STF's movement *away* from science—and even from "facts." As its name indicated, the Anti-Saloon League targeted the sale of alcohol rather than the substance itself. Although the League had recently endorsed national prohibition, its members were not even required to take the pledge against drink. So strategists were loath to sponsor campaigns about the physiological effect of liquor, focusing instead upon its social consequences—or simply forsaking argument altogether. Of fifty-one stereopticon slides that the League helped the STF produce, for example, just two addressed the dangers of drink to the individual. Several others addressed public perils like crime and accidents. But the vast majority depicted pithy statements by "famous men," ranging from Lloyd George and Lord Kitchener to Abraham Lincoln and Ty Cobb. These were "advertisements," Cora Stoddard admitted, aimed less at informing the public about alcohol than at provoking an immediate, emotional response from the "product audience." Born as a bulwark of science and knowledge, the Federation had forsaken both in order to stay afloat. Only when Stoddard reunited with the WCTU would she rediscover the original, physiological focus of Scientific Temperance.[56]

Rapprochement, 1917-1922

Several factors combined to induce a rapprochement between Cora Stoddard and the Woman's Christian Temperance Union in 1917. First of all, Edith Davis had fallen ill; the following spring, she would be dead. Upon America's entry into World War One, meanwhile, Stoddard was named secretary of the military's highly publicized Committee on War Temperance Activities. As more and more states ratified national prohibition, finally, the widespread persistence of drink in dry states demonstrated the ongoing need for instruction about its perils. "When you have once convinced an intelligent human being that alcohol in any form or quantity is a deadly, protoplasmic poison," the *Union Signal* editorialized in January 1918, "he does not attempt to cavil or indulge in sophistries on personal liberty, revenue, or states rights." Nor, the *Signal* added, does he ingest the substance himself. The same editorial announced the return of "scientific expert" Cora Frances Stoddard, who would help persuade Americans to close the "rum holes" under their noses as well as the saloons in their neighborhoods.[57]

Significantly, Stoddard came back as superintendent of the WCTU's Bureau of Scientific Temperance *Investigation* rather than its Department of Scientific Instruction. Here she would oversee data collection and "the education of the general public in the facts about alcohol," exactly as she did through her post at the STF. For the next few years, then, Stoddard helped the WCTU produce billboards, pamphlets, and posters reminding Americans about the inherent dangers of drink. Meanwhile, she also made a special effort to attract press notice for STI. "[T]here are thousands of people . . . to whom [the press] is law and gospel," Stoddard told an associate. "This may seem ridiculous, but it is a fact." A single favorable editorial was "worth more than many thousand small leaflets distributed at random," she added, because "leaders of opinion"—including educators—would listen.[58]

Especially after the war ended, however, schools across the country ignored Scientific Temperance Instruction. Stunned by reports that nearly 40 percent of war draftees had been rejected due to "physical and mental defects," educators instituted a revitalized health code stressing diet, exercise, and personal hygiene. Although textbooks showed a slight increase in the amount of space they devoted to alcohol, most teachers downplayed it in favor of other subjects. Students, for their part, seemed pleased to dispense with STI altogether. In an informal survey, pupils at a Chicago high school listed sex education, disease prevention, and sanitation as the health topics of greatest "practical value." Next came diet, digestion, "special senses," and care of the heart. Alcohol did not even make the list. "I find that our 'temperance' teaching . . . is being dropped to a large extent, and the new Health Code put in its place," reported Elizabeth Middleton, who had replaced Edith Davis as superintendent of Scientific Temperance Instruction. "We must guard against this." At Middleton's request, Stoddard began to craft a new course of study "showing how [STI] can be linked and interwoven with the other health topics."[59]

Yet she also began to realize that schools would never adopt this "integrated" curriculum unless "it were pressed upon them by citizens, courteously and understandingly." Rather than simply responding to hygienists in print, in other words, Stoddard now called upon the WCTU to rebuild the wide network of local STI assistants that had waned since the days of her mentor, Mary Hunt. In 1921 and 1922, she noted, WCTU women visited schools only to supervise annual essay contests and Temperance Day celebrations. Yet "essay work never reaches more than a small proportion of pupils," she argued, while "the temperance day program is but one day in the year"; neither could substitute for the "systematic instruction of *all* pupils." Meanwhile, most school boards refused to buy the books that "regular" STI required. Even where texts were purchased, they rarely found their way into pupils' hands. "[S]omehow from some quarter," Stoddard wrote WCTU president Anna Gordon in July 1922, "renewed pressure must be brought to bear upon the responsible officials—the old story, but just as important as ever, I know you will agree."[60]

Under the division of labor that the WCTU had devised back in 1918, however, the responsibility for such pressure lay with "Lizzie" Middleton and the STI department rather than with Stoddard's Bureau of Scientific Temperance Investigation. A brief power struggle ensued, as Stoddard demanded control over both divisions and Middleton defended the small gains she had made. "[T]he more important consideration is to have the regular text book teaching in the schools," Middleton admitted. "But I have found in so many schools that the text book teaching is so lax, frequently left out altogether, that about all the teaching the children get is in the essay contests and the Temperance Day observance." Regular instruction, in short, was desirable but hardly viable; if WCTU women called for it, they risked closing the only venue that was still open to them. Yet the votes were against Lizzie Middleton, or so it seemed. Under the pretense of ill health, she relinquished her position to Cora Stoddard at the Union's national convention in December 1922. For the first time since the death of Mary Hunt, STI lay under the direction of her best-known disciple.[61]

Requiem, 1922-1925

Mimicking Hunt almost word for word, Stoddard in her first "plan of work" called upon all STI workers to stream back into their schools—but to use caution when they got there. "[L]et the point of view be that of cooperation—we want to know what is being done so that we can help most intelligently," she advised. Visitors should pay special attention to the health code and other hygiene curricula, making sure they did not neglect alcohol. Here WCTU women could refer to Stoddard's "Outlines for Health and Temperance Teaching," which provided suggestions for how schools might profitably combine them. Most of all, Stoddard emphasized, every city, town, and village WCTU must assign one of its own to monitor STI. "The schools are *local*," she wrote, putting the last word in boldface for good measure. "A state superintendent can not do all that can or should be done to hold and foster temperance education in *your* schools."[62]

To Stoddard's dismay, however, few WCTU women at any level seemed to know—or care—about Scientific Temperance. Much of the problem stemmed from the advent of national prohibition, which appeared to render STI superfluous. Like Mary Hunt before her, Stoddard had long claimed that school lessons about the dangers of alcohol would cause citizens to vote for its ban. With the passage of the Eighteenth Amendment, then, "many seem to feel that Our work is done," as a state STI leader observed. Increasingly, women turned to so-called law enforcement activities like the formation of teenage National Prohibition Guards. Sporting bright banners and uniforms, these WCTU youth legions epitomized the ebullient Gospel Temperance tradition that STI had sought to supplant. For in their loud paeans to abstinence and obedience, the Prohibition Guards omitted any *argument* for either. "There is a tendency at the present time to emphasize the *must* of our prohibition law at the expense of the *why*," complained one Stoddard aide in 1923. "Unless the two go hand in hand there is bound to be rebellion, because those who do not understand the reasons will think it unjust and oppose it."[63]

For the most part, these warnings fell on deaf ears. State WCTUs frequently assigned their weakest candidates to supervise Scientific Temperance, if they appointed anyone at all. In Georgia, for example, the state WCTU convention named Mrs. Howard Park STI director even though she had "no previous training or knowledge" in the subject. "I had an interested expression on my face," Park explained, in a sheepish letter to Cora Stoddard. "If I had been put in the Congo and told to teach Temperance, I could not have been more ignorant of how to go about it than I have been with this work." Worst of all, even energetic state STI leaders proved unable to interest their constituents in the subject. Assuming New Hampshire's STI post in 1923, Anna E. Muzzey found that none of her county WCTUs had selected an STI superintendent. A subsequent plea to do so produced exactly one response. "The work of State Superintendent of STI is too big for me I guess," Muzzey mused. "You see the work had been discontinued for some years, and you know what that means." The constant drudgery of the job—studying textbooks, writing letters, and visiting schools—made women reluctant to volunteer for it.[64]

Finally, the few WCTU women who continued to tout STI met sharp resistance from teachers. Since so many instructors were female, especially at the primary level, several state WCTUs strove to enlist them as members. Tennessee's union established a "Win Every Teacher Week," hosting receptions and luncheons in schools across the state; in Missouri, meanwhile, county STI superintendents earned prizes by recruiting ten or more classroom instructors for the WCTU. Yet even teachers who joined the union sometimes neglected STI, thanks to the maelstrom of other curricular demands that swirled around them. "*If* we could have a supervisor of [STI] as the schools have supervisors of almost everything else we might have our work given its proper place," Middleton told Cora Stoddard. "But so long as other supervisors are urging and demanding so much from the teacher she cannot be blamed for letting this teaching slide aside."[65]

In the end, then, STI's key stumbling block remained where it had always been: with professional educators. "Without being unkindly critical," Stoddard wrote, "it does seem to me that our educational people . . . have shown for 40 years an appalling amount of indifference and incapability in this matter of temperance instruction." When STI began, she noted, "fear of political pressure"—especially from local ward bosses—often dissuaded schoolmen from enforcing it. Now that direct popular control had declined, however, educators' own professional concerns got in the way. In 1923, for example, the NEA rejected STI as insufficiently "expert" to include on the program of its annual convention. "The National Education Association must approach its problems from the standpoint of the professional educator," NEA president William B. Owen told an aide to Cora Stoddard. "[W]e are not organized to carry on general movements outside our own work as school teachers."[66]

Yet as Stoddard realized, large fractions of the American curriculum reflected "general movements" beyond the classroom doors. "[E]very new subject which has gone into the schools, has had to be started and pressed from the outside," she noted in 1924. "[I]t seems to me a bit appalling that our own women should fail to avail themselves of this opportunity."[67] As she wrote, indeed, hundreds of other citizen groups were converging upon state legislatures, school boards, and principals' offices to influence public school curricula. Patriotic societies like the Sons of the American Revolution demanded greater attention to the Founding Fathers in history courses, while ethnic groups pleaded for more instruction about their own contributions to the national past. Confederate veterans appealed for textbooks that celebrated Southern valor in the "War Between the States," eliciting howls of protest from Northern military societies. Both groups supported school drill practice, while pacifists demanded physical education in its place. Nativist organizations won periodic bans on foreign-language instruction and so-called sectarian indoctrination. Religious groups, meanwhile, pressed for Bible reading in schools and—most notoriously—for restrictions on the teaching of evolution.[68]

Yet none of these efforts was as ubiquitous as Scientific Temperance Instruction, surely the most successful lay movement in American educational history.[69] More than any other popular struggle, then, Scientific Temperance should also help us decide what sort—and what amount—of "lay influence" we *want* in our schools. Peaking at precisely the moment of STI's final demise, the antievolution movement calls our attention to the repressive underside of citizen control. As Scientific Temperance suggests, however, even citizens who seek to silence public conversation may—in practice— spark it. Despite the despotic and single-minded impulses of its founder, Mary Hanchett Hunt, STI would engage hundreds of thousands of Americans in a compelling dialogue about alcohol, expertise, and democracy— without "compelling belief" in anyone.[70]

EPILOGUE

dis-till vb: 1. to extract the essence of; to concentrate 2. to purge the volatile elements from

In 1902, school superintendent W. B. Ferguson of Middletown, Connecticut, published a lengthy attack upon Mary Hunt and Scientific Temperance Instruction. Echoing chemist Wilbur O. Atwater, a fellow Middletown resident, Ferguson mocked Hunt's claims that alcohol was poisonous. Most of all, though, Ferguson reviled STI's "old Socratic philosophy" that "knowledge of evil insures the avoidance of evil." If anything, he insisted, STI's emphasis upon the harms of alcohol would promote the very sin it sought to prevent. "We do not teach hygiene by the study of disease," Ferguson wrote, quoting Teachers College professor Samuel T. Dutton, "nor purity by the contemplation of vice."

Nevertheless, Ferguson could take comfort that history was on his side. "The twentieth century," he wrote the previous year,

> will doubtless witness marvelous progress in many lines of effort. That progress will be made, as progress always has been made, through intense thought and study rather than through mere sentiment and feeling. I do not undervalue the work sometimes accomplished by enthusiasts . . . but they need to be directed and controlled, else they are like wild engines, whose speed is dangerous and whose destination is uncertain. Clear-headed, practical men and women must direct any reform, if the results are to be beneficial and lasting.[1]

This book has tried to shed new light upon the complex relationship between "enthusiasts" and "practical men and women"—that is, laypeople and professionals—during the Progressive era. Mimicking Ferguson, the first historians of this period argued that experts defined and defended an often unruly, undisciplined popular will. Scholars in the 1960s and 1970s reversed this judgment, claiming that experts used new knowledge and power to subjugate, not elucidate, lay voices. Most recently, historians have singled out two intersecting sets of experts—female reformers and liberal intellectuals—who sought to reconcile popular and professional authority. Democrats rather than technocrats, figures like Jane Addams and John Dewey envisioned expert knowledge circulating through an active, informed lay polity.

How did laypeople respond to this new knowledge? Focusing almost solely upon the experts themselves, most historical accounts leave lay citizens out of the picture. At least implicitly, we tend to presume that laypeople were confused, alienated, muzzled, or bedazzled by the modern sirens of professional power. In the case of Scientific Temperance, however, Hunt and her WCTU legions embraced expertise because they understood it—especially its remarkable potential to advance their agenda. Just as W. B. Ferguson cited Wilbur O. Atwater and Samuel T. Dutton to condemn STI, Mary Hunt could seize upon Winfield S. Hall and William T. Harris to support it. Scientific Temperance reveals how *laypeople* could reconcile professional and popular authority, suggesting a fresh way to understand the overall boom in Progressive-era expertise.

Second, STI shows how one group of lay Americans fused modern expertise to a more traditional faith in personal autonomy. Scholars usually associate this ethos with the "old," or nineteenth-century, middle class, which coalesced around concepts of inner discipline and self-control. In the twentieth century, the story goes, a "new" middle class stressed the social roots of behavior: sin and vice stemmed from degraded environments, not from depraved souls. For Mary Hunt and her assistants, though, the independent individual still reigned supreme. Hunt dressed this faith in the glossy rhetoric of laboratory science, but its bedrock philosophy remained the same: if properly informed, individuals would use their powers of reason to render the right choice. Neither "traditional" nor "modern," then, STI combined elements from each world. Retaining older premises about free will and agency, it adopted new forms of knowledge to buttress and propagate them.

Third, the struggle over Scientific Temperance suggests a different approach to the study of Progressive-era school politics. Most historians still presume that educational officials dominated the American curriculum during this period.[2] Yet a vast number of citizen initiatives dotted the course of study, too numerous to detail here. The Grand Army of the Republic lobbied for military training in the schools; the Ku Klux Klan, for Bible reading; and the United Confederate Veterans, for "a proper and truthful history" of "the War Between the States." Other groups were best known for prohibitions, not prescriptions. The Committee on Militarism in Education pressed legislatures to ban drill instruction; the American Defense Society, to restrict German; and the World's Christian Fundamentals Association, to bar the teaching of evolution. Like the WCTU, moreover, each of these organizations sent members into the schools to enforce its will. Hardly a lone citizen bulwark on a sea of expert edicts, then, STI reflected a pattern of lay curricular influence that would confound professional educators across our entire century.[3]

Into the present, finally, Scientific Temperance highlights a sharp tension within liberal theories on democracy and education. In public schools, liberals hoped, children would learn the essential skills of citizenship: argument, analysis, communication, and compromise. As the STI saga shows, though, liberals tried to insulate schools from the same deliberative processes that education was supposed to inculcate. Even John Dewey, America's most eloquent prophet of the democratic faith, did not believe that lay citizens could set the proper contours of school policy. "Are the schools doing what the people want them to do?" he asked rhetorically in 1901. "The schools are not doing, and cannot do, what the people want until there is more unity, more definiteness, in the community's consciousness of its own needs; but it is the business of the school to forward this conception."[4] The public, in Dewey's scheme, would not deliberate the content or purpose of school. Instead, the school—like the egg before the chicken—would hatch a deliberative public.

When alcohol is distilled, its essence is extracted; at the same time, volatile elements are purged. Distilling democracy, liberals like Dewey hoped to extract its essence: an informed, critically minded public. Simultaneously, though, they also sought to purge volatile elements—that is, lay voices—from public school policy making. Noting the "popular pressure and clamor" for new school subjects, Dewey flatly declared that experts should select them. "An enlightenment of vision is the prerequisite of efficiency in conduct," he wrote. "[W]hen the educators have come to some agreement as to what education is, the community will . . . make their ideal a reality."[5] Trained professionals would dictate curricular policy, in sharp contrast to the theories of knowledge and democracy that Dewey expounded elsewhere.

In his analyses of school administration, for example, Dewey condemned autocratic restrictions upon classroom teachers. "What does democracy mean save that the individual is to have a share in determining the conditions and the aims of his own work?" Dewey asked. "How can we justify our belief in the democratic principle elsewhere, and then go back entirely upon it when we come to education?" The same argument would suggest that school patrons must "share in determining" educational policy, too, lest it negate Dewey's other paeans to popular rule. For Dewey, however, education played roughly the same role as revolution did for Karl Marx: it was the precondition of progress, so only a vanguard could deliver it. Until education was properly established, moreover, most popular discussion of the subject could be safely dismissed as false consciousness. Rather than capitulating to "the meaningless and arbitrary flux and reflux of public sentiment," Dewey concluded, schools should educate the public so its sentiment made sense.[6]

As the struggle over Scientific Temperance demonstrates, though, school politics can itself become a venue for informed, impassioned citizen debate. Thousands of school boards held public hearings on STI, so continual—and contentious—that Mary Hunt organized mock sessions to prepare for them. Scores of WCTU women streamed into classrooms, assisting as well as antagonizing teachers. In the daily press, meanwhile, Hunt and her supporters exchanged fiery words with Nicholas Murray Butler, David Starr Jordan, and other eminent educators. And within the WCTU itself, finally, national leaders reviled Scientific Temperance, while local members rallied to its defense. Reviewing this jumble of discord in 1902, a dejected Mary Hunt told a friend that "the storms beat about my head." That same year, however, at least one local STI worker could see a rainbow above them. "It has set the people to *thinking* and *talking*, at least," wrote Mary Donaldson from tiny Ellenville, New York. Surely, she concluded, "that was something."[7]

But to John Dewey, of course, it was everything. His central goal was to improve what he called "the conditions of communication," to get more people "thinking and talking" about their common goals—and their common lives. Educational policy would seem uniquely suited for such a dialogue, in fact, because schools still represent our chief public vehicle for defining and disseminating our values, principles, and priorities. So rather than condemning lay curricular initiatives out of hand, as Dewey did, we must examine each one to see *whether* it promoted discussion. Openly racist or sexist curricula would fail this test, because by their very essence they restrict deliberation. Provided that dissent is protected, though, we should encourage continual citizen participation in shaping what schools teach. Let a thousand curricular flowers bloom: under the proper conditions, they can nurture the democracy that sows them.[8]

Consider, once more, the issue that initially sparked this book: contemporary drug and alcohol education. More than three-quarters of the nation's school districts have now adopted a single antidrug program, Drug Abuse

Resistance Education (DARE), making it the most ubiquitous American health curriculum since Scientific Temperance Instruction. Like STI, DARE stresses total abstinence. Just as STI pupils a century ago frequently drank alcohol, however, children today take DARE *and* drugs. Nearly every controlled study of DARE during the past decade has concluded that it does not decrease students' use of illegal narcotics. Indeed, several investigators found DARE graduates more likely to use certain drugs than children who had not received this training. Citing these studies, DARE's critics argue that expert-minted knowledge—not "popular pressure and clamor," as Dewey called it—should determine drug education. "Vast resources of money, personnel, time, and effort are being expended on DARE and other ineffective no-use programs," one scholar concluded in 1996, after reviewing a massive array of research. All future curricula "should be based on science and objective evidence rather than on political ideology," he added.[9]

Under the theory sketched here, by contrast, curricula should reflect the will—even the "ideology"—of school patrons, whether "science" supports them or not. (The scientists have their own ideology, of course, but that is another story.) Even DARE's harshest critics acknowledge that the program is wildly popular among students, parents, and community leaders. How can these citizens receive the necessary tutelage in democracy if the very institution that tutors them—the public school—is itself undemocratic? The current DARE regime teaches children that all illegal drug use is harmful, an almost laughably false proposition. But a "realistic" or "objective" curriculum, fashioned according to expert dictates, would teach them something much worse: that their voices do not matter. Our drug education might improve, but our democracy would be impoverished.

At the same time, though, we must also ensure that popular curriculum efforts do not themselves quell public discussion and deliberation. American history is replete with such movements, from Protestants who torched Catholic homes during antebellum battles over Bible instruction to nativists who burned German-language books during World War One. Most notoriously, the antievolution struggle of the 1920s led to a sharp narrowing of opinion: not only were textbooks purged of "Darwinist" doctrine, but teachers were dismissed for instructing it.[10] Likewise, recent evidence suggests that DARE organizers have silenced certain opponents through a potent mix of public relations and behind-the-scenes intimidation. Critics of DARE often request to remain anonymous, for instance, a strong sign that dissent is being muzzled.[11]

Here Scientific Temperance stands as a massive counterexample, an enduring reminder that popular curricula can transmit majority beliefs without trampling upon minority ones. In seven years of research, I failed to find even one instance of a teacher who was fired for criticizing STI. Nor did I locate a single student who was expelled, suspended, or otherwise disciplined for doing so. Finally, I found no evidence that Mary Hunt "scared away writers and publishers interested in preparing books that would be real alternatives to the books she promoted," as one historian has charged.[12] On the contrary—and much to Hunt's dismay—competing books continued to appear. In some states, indeed, these so-called wet texts proved more popular than the books Hunt supported.

For one final time, I must emphasize that Mary Hunt and her WCTU assistants never sought to elicit deliberation or discussion in American schools. Yet intentions and outcomes often diverge, as Hunt and I would each discover. I had aimed to write a book that skewered Scientific Temperance, but I ended up endorsing its democratic achievement. Hunt sought to stifle this same democratic quality, but she ultimately—and unwittingly—enhanced it. Now that the book is complete, my only hope is that it sparks more debate about schools, citizens, and expertise. A faith in democracy demands nothing less.

NOTES

PREFACE

1. I expanded this argument in "Amotivation and Other Syndromes," *Baltimore Sun*, 14 June 1991. Lest this article be mistaken for an act of bravery, I should emphasize that I wrote it after leaving high-school teaching—not before.

2. Mary H. Hunt, *An Epoch of the Nineteenth Century: An Outline of the Work for Scientific Temperance Education in the Public Schools of the United States* (Boston: P. H. Foster and Co., 1897), 47.

3. This is the theme of David Tyack, et al., *Law and the Shaping of Public Education, 1785-1954* (Madison: University of Wisconsin Press, 1987), ch. 6, which links Scientific Temperance to antievolution and patriotism movements in the schools. Other critical examinations of STI include Howard K. Beale, *A History of Freedom of Teaching in American Schools* (New York: Charles Scribner's Sons, 1941), ch. 6; Richard K. Means, *A History of Health Education in the United States* (Philadelphia: Lea and Febiger, 1962), 50-56; James H. Timberlake, *Prohibition and the Progressive Movement, 1900-1920* (Cambridge: Harvard University Press, 1963), 48-50; Philip J. Pauly, "The Struggle for Ignorance about Alcohol: American Physiologists, Wilbur Olin Atwater, and the Woman's Christian Temperance Union," *Bulletin of the History of Medicine* 64 (1990): 366-92; and Alice Yageric, "Instruction in the Effects of Alcohol and Narcotics in New York State Public Schools, 1884-1933" (Ph.D. diss., State University of New York at Buffalo, 1995), chs. 4-5. More sympathetic accounts of STI appear in Norton Mezvinsky, "The White-Ribbon Reform, 1874-1920" (Ph.D. diss., University of Wisconsin, 1959), ch. 7; idem, "Scientific Temperance Instruction in the Schools," *History of Education Quarterly* 1 (Spring 1961): 48-56; Ruth Bordin, *Woman and Temperance: The Quest for Power and Liberty, 1873-1900* (Philadelphia: Temple University Press, 1981), 135-38; and Jonathan Zimmerman, "'The Queen of the Lobby': Mary Hunt, Scientific Temperance, and the Dilemma of Democratic Education in America, 1879-1906," *History of Education Quarterly* 32 (Spring 1992): 1-30.

4. F. C. Atwell to R. H. Magwood, 5 April 1904, frames 739-42, roll 12, Temperance and Prohibition Papers, Ohio Historical Society (joint Ohio Historical Society—Michigan Historical Collections), Scientific Temperance Federation Series [hereafter STF Series].

5. *Educational Review* 11 (1896): 307.

6. See, e.g., Charles Frank Speierl Jr., "Influence of the Grand Army of the

Republic on Education in New Jersey between 1865-1935" (Ed.D. diss., Fairleigh Dickinson University, 1987); Stuart McConnell, *Glorious Contentment: The Grand Army of the Republic, 1865-1900* (Chapel Hill: University of North Carolina Press, 1992), 224-32; Samuel J. McLaughlin, "The Educational Policies & Activities of the American Federation of Labor during the Present Century" (Ed.D. diss., New York University 1936), 21.

7. As David Tyack and his collaborators recently concluded, "The WCTU was perhaps the most influential lay lobby ever to shape what was taught in public school." Tyack, et al., *Law and the Shaping of Public Education*, 161.

I. INTRODUCTION

1. Mary H. Hunt, *A History of the First Decade of the Department of Scientific Temperance Instruction . . .* , 2nd ed. (Boston: George E. Crosby, 1891), 80.

2. C. S. Winchell, "Report of the State Superintendent of Scientific Temperance Instruction," *Report of the Eleventh Annual Convention of the Woman's Christian Temperance Union of the State of Minnesota* (Red Wing, Minn.: Red Wing Printing Co., 1887), 77-78; H. A. Hobart, "Annual Address of the President," ibid., 71; "Discussion," *Journal of Proceedings and Addresses of the National Education Association* 39 (1900): 256; C. K. G. Hyde to Mary Hanchett Hunt [hereafter MHH], 3 June 1887, frames 423-24, roll 15, Temperance and Prohibition Papers, Ohio Historical Society (joint Ohio Historical Society—Michigan Historical Collections), Scientific Temperance Federation Series [hereafter STF Series].

3. MHH to Hyde, 7 June 1887, frame 427, roll 15, STF Series; L. F. Whitaker, "To the White Ribboners of Washington County," *Union Signal*, 6 Oct. 1887; Mary H. Hunt, *An Epoch of the Nineteenth Century: An Outline of the Work for Scientific Temperance Education in the Public Schools of the United States* (Boston: P. H. Foster, 1897), 46. For a list of the signatories to the petition, see *Minutes of the Fourteenth Convention of the National Woman's Christian Temperance Union* (Chicago: Woman's Temperance Publishing Co., 1887), cxli-cxliv.

4. See, e.g., JoAnne Brown, *The Definition of a Profession: The Authority of Metaphor in the History of Intelligence Testing, 1890-1930* (Princeton: Princeton University Press, 1992); Thomas L. Haskell, *The Emergence of Professional Social Science: The American Social Science Association and the Nineteenth-Century Crisis of Authority* (Urbana: University of Illinois Press, 1977); Thomas L. Haskell, ed., *The Authority of Experts: Studies in History and Theory* (Bloomington: Indiana University Press, 1984); James T. Kloppenberg, *Uncertain Victory: Social Democracy and Progressivism in European and American Thought, 1870-1920* (New York: Oxford University Press, 1986); James Marone, *The Democratic Wish: Popular Participation and the Limits of American Government* (New York: Basic, 1990), ch. 3; Alexandra Oleson and John Voss, eds., *The Organization of Knowledge in Modern America, 1860-1920* (Baltimore: Johns Hopkins University Press, 1979); Dorothy Ross, *The Origins of American Social Science* (Cambridge, Eng.: Cambridge University Press, 1991); Robert Wiebe, *Self-Rule: A Cultural History of American Democracy* (Chicago: University of Chicago Press, 1995), pt. 2; Laura M. Westhoff, "The Popularization of Knowledge: John Dewey on Experts and American Democracy," *History of Education Quarterly* 35 (Spring 1995): 27-47.

5. The classic formulation of this bipolar model was Richard Hofstadter's Pulitzer-Prize-winning study, *Anti-Intellectualism in American Life*. To Hofstadter, intellectual and popular authority were locked into a bitter struggle that flared intermittently across the entire national past. During the Progressive era, it lay dormant: ascendant

experts, Hofstadter wrote, "were widely accepted by the public." But it exploded once again in the 1920s, when a rejuvenated village Protestantism faced off against cosmopolitan challenges like Darwin and Freud. More recently, Nathan Hatch edited a set of essays stressing America's "veritable love-hate relationship with the role of the expert," as Hatch wrote. For the past century, contributor Laurence Veysey added, "bureaucrats, technocrats, and professionals have grown accustomed . . . to being placed on the defensive, after an intervening period when everything seemed to be going their way." Hofstadter, *Anti-Intellectualism in American Life* (New York: Vintage, 1962), 205; Hatch, "Introduction: The Professions in a Democratic Culture," in *The Professions in American History* (South Bend, Ind.: University of Notre Dame, 1988), 2; Veysey, "Higher Education as a Profession: Changes and Continuities," ibid., 23.

6. See, e.g., Paul E. Johnson, *A Shopkeeper's Millennium: Society and Revivals in Rochester, New York, 1815-1837* (New York: Hill and Wang, 1978); Mary P. Ryan, *Cradle of the Middle Class: The Family in Oneida County, New York, 1790-1865* (Cambridge, Eng.: Cambridge University Press, 1981); Stuart Blumin, *The Emergence of the Middle Class: Social Experience in the American City, 1760-1900* (Cambridge, Eng.: Cambridge University Press, 1989), ch. 6; Daniel Walker Howe, "The Evangelical Movement and Political Culture in the North during the Second Party System," *Journal of American History* 77 (1991): 1216-39.

7. On the mounting "social" consensus for prohibition during these years, the classic text is James H. Timberlake, *Prohibition and the Progressive Movement, 1900-1920* (Cambridge, Mass.: Harvard University Press, 1963). More recent statements include Norman H. Clark, *Deliver Us From Evil: An Interpretation of American Prohibition* (New York: W. W. Norton, 1976); K. Austin Kerr, *Organized for Prohibition: A New History of the Anti-Saloon League* (New Haven: Yale University Press, 1985); Richard F. Hamm, *Shaping the Eighteenth Amendment: Temperance Reform, Legal Culture, and the Polity, 1880-1920* (Chapel Hill: University of North Carolina Press, 1995); Catherine Gilbert Murdock, "Domesticating Drink: Women and Alcohol in Prohibition America, 1870-1940" (Ph.D. diss., University of Pennsylvania, 1995).

8. These observations are drawn from a burgeoning literature, summarized in Michael Kazin, "The Grass-Roots Right: New Histories of U.S. Conservatism in the Twentieth Century," *American Historical Review* 97 (1992): 136-55; Alan Brinkley, "The Problem of American Conservatism," *American Historical Review* 99 (1994): 409-29; Leo P. Ribuffo, "Why Is There So Much Conservatism in the United States and Why Do So Few Historians Know Anything about It?" *American Historical Review* 99 (1994): 439-49; Leonard J. Moore, "Good Old-Fashioned New Social History and the Twentieth-Century American Right," *Reviews in American History* 24 (1996): 555-73. My emphasis upon conservatives' individualist ethos echoes Susan M. Yohn, who distinguishes conservatives by their "focus on both individual morality and the process of conversion" and their "suggestion that social problems were the result of sinful and immoral behavior on the part of individuals." Yohn, "Will the Real Conservative Please Stand Up? or, the Pitfalls Involved in Examining Ideological Sympathies: A Comment on Alan Brinkley's 'Problem of American Conservatism,'" *American Historical Review* 99 (1994): 433.

9. Frederick Jackson Turner, "Pioneer Ideals and the State University," in idem, *The Frontier in American History* (New York: Henry Holt, 1920), 285.

10. One notable exception was Samuel P. Hays's *Conservation and the Gospel of Efficiency: The Progressive Conservation Movement, 1890-1920* (Cambridge: Harvard University Press, 1959), which examined how Gifford Pinchot and other foresters used Progressive reform to enhance — indeed, to construct — their own professional authority.

So far as I know, however, no other historian before the 1960s cast such a skeptical eye on experts and expert power.

11. See, e.g., Gabriel Kolko, *The Triumph of Conservatism: A Reinterpretation of American History, 1900-1916* (New York: Free Press, 1963); James Weinstein, *The Corporate Ideal in the Liberal State, 1900-1918* (Boston: Beacon, 1968). Later works in this tradition include David F. Noble, *America By Design: Science, Technology, and the Rise of Corporate Capital* (New York: Knopf, 1979); idem, *Forces of Production: A Social History of Industrial Automation* (New York: Oxford University Press, 1984); and Magali Sarfatti Larson, *The Rise of Professionalism* (Berkeley: University of California Press, 1977).

12. Robert Wiebe, *The Search for Order, 1877-1920* (New York: Hill and Wang, 1967), 174. Other representative works included Samuel Haber, *Efficiency and Management: Scientific Management in the Progressive Era, 1890-1920* (Chicago: University of Chicago Press, 1964); Jerry Israel, ed., *Building the Organizational Society: Essays on Associational Activity in Modern America* (New York: Free Press, 1972); and David Tyack, *The One Best System: A History of American Urban Education* (Cambridge: Harvard University Press, 1974).

13. On Wiebe's use of modernization theory, see Kenneth Cmiel, "Destiny and Amnesia: The Vision of Modernity in Robert Wiebe's *The Search for Order,*" *Reviews in American History* 21 (1993): 352-68.

14. See, e.g., Jerome Auerbach, *Unequal Justice: Lawyers and Social Change in Modern America* (New York: Oxford University Press, 1976) and especially Burton J. Bledstein, *The Culture of Professionalism: The Middle Class and the Development of Higher Education in America* (New York: Norton, 1976).

15. Often, to be sure, this outcome was unintended. In New York, Richard McCormick found, "popular outrage" against corruption spawned new agencies to regulate railroads, insurance companies, and other corporations. Yet these reforms soon "weakened the old patterns of participation," replacing the white-hot spirit of democratic politics with the drab efficiency of expert administration. Irony and pessimism pervaded McCormick's interpretation. A genuinely popular effort to restore the grass roots of government, Progressive-era reform ended up empowering the experts instead. Richard McCormick, *From Realignment to Reform: Political Change in New York State, 1893-1910* (Ithaca, N.Y.: Cornell University Press, 1981), 268-69.

16. Theda Skocpol and Gretchen Ritter, "Gender and the Origins of Modern Social Policies in Britain and the United States," *Studies in American Political Development* 5 (Spring 1991): 84; Robyn Muncy, *Building a Female Dominion in American Reform, 1890-1935* (New York: Oxford University Press, 1991), 59; Theda Skocpol, *Protecting Soldiers and Mothers: The Political Origins of Social Policy in the United States* (Cambridge: Harvard University Press, 1991), 494.

17. Thomas Bender, "Academic Knowledge and Political Democracy in the Age of the University," in *Intellect and Public Life: Essays on the Social History of Academic Intellectuals in the United States* (Baltimore: Johns Hopkins University, 1993), 132; Hilary Putnam, "A Reconsideration of Deweyan Democracy," in *Pragmatism, Law and Society,* ed. Michael Brint and William Weaver (Boulder, Colo.: Westview Press, 1991), 217; Westhoff, "The Popularization of Knowledge," 32-34; John Dewey, *The Public and its Problems* (Denver: Henry Holt, 1927), 208. On the revival of pragmatism, see James T. Kloppenberg, "Pragmatism: An Old Name for Some New Ways of Thinking?" *Journal of American History* 83 (June 1996): 100-138.

18. Willa Cather, *Not Under Forty* (New York: Knopf, 1936), v. For examples of historians' use of this remark, see Warren Susman, *Culture as History: The Transformation of American Society in the Twentieth Century* (New York: Pantheon, 1984),

105; John C. Burnham, *Bad Habits: Drinking, Smoking, Taking Drugs, Gambling, Sexual Misbehavior, and Swearing in American History* (New York: New York University Press, 1993), 3; Lynn Dumenil, *The Modern Temper: American Culture and Society in the 1920s* (New York: Hill and Wang, 1995), 3.

19. Hofstadter, *Anti-Intellectualism in American Life,* 42–43; idem, *The Age of Reform* (New York: Vintage, 1955), 289–90; Wiebe, *Self-Rule,* 185; Moore, "Good Old-Fashioned New Social History," 561.

20. George M. Marsden, *Understanding Fundamentalism and Evangelicalism* (Grand Rapids, Mich.: William B. Eerdmans, 1991), 3; Wilfred McClay, *The Masterless: Self and Society in Modern America* (Chapel Hill: University of North Carolina Press, 1994), 154; Ferenc Morton Szasz, *The Divided Mind of Protestant America, 1880-1930* (University: University of Alabama Press, 1982), 44; idem, "The Progressive Clergy and the Kingdom of God," *Mid-America* 55 (1973): 16–17; Susan Curtis, *A Consuming Faith: The Social Gospel in Modern American Culture* (Baltimore: Johns Hopkins University Press, 1991), 5.

21. Marsden, *Understanding Fundamentalism,* 30.

22. George M. Marsden, "Evangelicalism and the Scientific Culture: an Overview," in *Religion and Twentieth-Century American Intellectual Life,* ed. Michael Lacey (Cambridge, Eng.: Cambridge University Press, 1989), 41; Szasz, *Divided Mind of Protestant America,* 64–65. By "individualist Right," I mean only that conservatives held each individual responsible for his or her actions. Likewise, I see liberals as "environmentalist" in the sense that they "placed the social formation of individual consciousness above the accountability of individuals for their own moral choices," as Richard Wightman Fox has written. I am *not* referring to their respective positions on the rights and freedoms of individuals within the social structure. Fox, "The Culture of Liberal Protestant Progressivism, 1875-1925," *Journal of Interdisciplinary History* 23 (1993): 655.

23. Contrary to most historical assumptions, millions of middle-class women as well as men used alcohol during these years. Indeed, women helped render drink "respectable" by developing a moderate, home-based style of consumption. Murdock, "Domesticating Drink," ch. 5, esp. 139–43.

24. Haskell, *Emergence of Professional Social Science,* 251. The classic account of antimodernism at the turn of the century is T. J. Jackson Lears, *No Place of Grace: Antimodernism and the Transformation of American Culture, 1880-1920* (New York: Pantheon, 1981).

25. Although few historians have described these Progressive-era conservatives, their strength can be seen in continued liberal attacks upon them. "How vain to expect better conditions," wrote sociologist E. A. Ross in 1907, in a typical denunciation, "simply by adding to the number of good men." Eight years later, Randolph Bourne derided the naive but omnipresent notion that "social ills may be cured by personal virtue." The very need to denounce this philosophy paid tribute to its ongoing vitality. David Danbom, *The World of Hope: Progressives and the Struggle for an Ethical Public Life* (Philadelphia: Temple University Press, 1987), 137; Lawrence B. Goodheart and Richard O. Curry, "A Confusion of Voices: The Crisis of Individualism in Twentieth-Century America," in *American Chameleon: Individualism in Trans-National Context,* ed. Curry and Goodheart (Kent, Ohio: Kent State University Press, 1991), 190.

26. The term "populist," especially, connotes "'ordinary' or 'average' men and women" who used "diligence, practical intelligence, and faith in God" to counter distant, dissipated elites. Kazin, "Grass-Roots Right," 141; idem, *The Populist Persuasion* (New York: Basic, 1995), 2. Hence historians routinely contrast "populist" to "expert,"

blinding us to grass-roots movements that harnessed expert authority to their own purposes. See, e.g., Veysey, "Higher Education as a Profession," 23; Gary Wills, *Under God: Religion and American Politics* (New York: Simon and Schuster, 1990), 111; Ronald Numbers, *The Creationists: The Evolution of Scientific Creationism* (New York: Knopf, 1992), 17; Moore, "Good Old-Fashioned New Social History," 560.

27. Charles H. Stowell, *A Primer of Health: Practical Hygiene for Pupils in Primary and Lower Grades* (New York: Silver, Burdett and Co., 1906), 67. The ensuing summary is based primarily upon the following textbooks, all endorsed by Mary Hunt and the Woman's Christian Temperance Union: Albert F. Blaisdell, *How to Keep Well: A Text-Book of Physiology and Hygiene for the Lower Grades of Schools,* rev. ed. (Boston: Ginn and Co., 1904); idem, *Our Bodies and How We Live: An Elementary Text-Book of Physiology and Hygiene for Use in Schools,* rev. ed. (Boston: Ginn and Co., 1904); idem, *Life and Health: A Text-Book on Physiology for High Schools, Academies, and Normal Schools* (Boston: Ginn and Co., 1902); Jeannette Winter Hall, *The New Century Primer of Hygiene for Fourth Year Pupils* (New York: American Book Co., 1901); Winfield S. Hall and Jeannette Winter Hall, *Intermediate Physiology and Hygiene for Lower Grammar Grades* (New York: American Book Co., 1901); Charles H. Stowell, *A Healthy Body: A Text-Book on Physiology and Hygiene, for Use in Intermediate Grammar Grades* (New York: Silver, Burdett and Co., 1906); idem, *The Essentials of Health: A Text-Book on Anatomy, Physiology, and Hygiene for Use in Higher Grades* (New York: Silver, Burdett and Co., 1906).

28. Blaisdell, *How to Keep Well,* 76.

29. Stowell, *Primer of Health,* 123.

30. Blaisdell, *Our Bodies and How We Live,* 97; Stowell, *A Healthy Body,* 28, 30.

31. Ray Ginger, *Six Days or Forever? Tennessee v. John Thomas Scopes* (Boston: Beacon, 1958), 44, quoted in Hofstadter, *Anti-Intellectualism,* 128n.

32. Hofstadter, *Anti-Intellectualism,* 407-8.

33. See, e.g., Szasz, *Divided Mind of Protestant America,* 132-33; William Sullivan, *Work and Integrity: The Crisis and the Promise of Professionalism in America* (New York: Harper Business, 1995), 94-95.

34. See, e.g., David Tyack, et al., *Law and the Shaping of Public Education, 1785-1954* (Madison: University of Wisconsin Press, 1987), 157-62; Alice Yageric, "Instruction in the Effects of Alcohol and Narcotics in New York State Public Schools, 1884-1933" (Ph.D. diss., State University of New York at Buffalo, 1995), chs. 4-5.

2. "A LITTLE BRIEF AUTHORITY"

1. Mrs. J. D. Weeks, "Report on Scientific Temperance," *Eleventh Annual Report of the Woman's Christian Temperance Union of Pennsylvania* (Pittsburgh: Anderson and Son, 1885), 63; Mary H. Hunt, *An Epoch of the Nineteenth Century: An Outline of the Work for Scientific Temperance Education in the Public Schools of the United States* (Boston: P. H. Foster, 1897), 9-11.

2. *Report of the Commissioner of Education for the Year 1886-1887* (Washington, D.C.: Government Printing Office [hereafter "GPO"], 216; "Minutes of the Twelfth Annual National Convention," *Union Signal,* 19 Nov. 1885; Anna M. Greene, "Pennsylvania: State Notes and Scientific Institutes," *Union Signal,* 28 May 1891; *Report of the Superintendent of Public Instruction of the Commonwealth of Pennsylvania* (Harrisburg: Edwin K. Myers, 1891), 128-29; Josephine Weeks to Mary H. Hunt [hereafter MHH], 15 April 1885, frame 366, roll 14, Temperance and Prohibition Papers, Ohio Historical Society (joint Ohio Historical Society—Michigan Historical Collections), Scientific Temperance Federation Series [hereafter STF Series].

3. Weeks to MHH, 8 Feb. 1885, frame 361, roll 14; Mary Lovell to MHH, 19 Aug. 1886, frame 202, roll 15, both in STF Series.

4. "Mrs. Mary H. Hunt," *Union Signal*, 3 June 1906; "Temperance: A Sketch of Mrs. Mary H. Hunt of Hyde Park," n.d., frame 273; Joseph Cook, "A Binary Star," n.d., frames 290-94; Edward Everett Hale, "A Bit of Modern Chivalry," 22 Oct. 1887, frame 322, all in roll 13, STF Series.

5. Robert H. Abzug, *Cosmos Crumbling: American Reform and the Religious Imagination* (New York: Oxford University Press, 1994), 48, 36; Keith Hardamann, *Charles Grandison Finney: Revivalist and Reformer* (Syracuse: Syracuse University Press, 1987), 126.

6. Jonathan Zimmerman, "Dethroning King Alcohol: The Washingtonians in Baltimore, 1840-1845," *Maryland Historical Magazine* 87 (Winter 1992): 380, 389-90; Robert Hampel, *Temperance and Prohibition in Massachusetts, 1813-1852* (Ann Arbor: UMI Research Press, 1982), 118; Milton A. Maxwell, "The Washingtonian Movement," *Quarterly Journal of Studies on Alcohol* 11 (1950): 412.

7. Hale, "A Bit of Modern Chivalry," frame 301; "Raised Without Rum," *Connecticut Western News* [Litchfield], 13 Feb. 1902, frame 500; manuscript by Emma Transeau Benedict, n.d. [hereafter "Transeau biography"], frame 511, all in roll 13, STF Series.

8. Helen Buss Mitchell, "'The North and South Here Meet': Almira Hart Lincoln Phelps and the Patapsco Female Institute, 1841-1856" (Ph.D. diss., University of Maryland, 1990), 7-13, 226.

9. Mitchell, "'The North and South Here Meet,'" 222; Emma Lydia Bolzau, *Almira Hart Lincoln Phelps: Her Life and Work* (Lancaster, Pa.: Science Press Printing Co., 1936), 396-97.

10. Transeau biography, frames 458-59, roll 13, STF Series; Cora Stoddard to E. O. Taylor, 31 Jan. 1907, frame 510, roll 45, STF Series; Sam Bass Warner, *Streetcar Suburbs: The Process of Growth in Boston, 1870-1900* (Cambridge: Harvard University Press, 1978), 71 (showing map of Hyde Park and other adjacent suburbs); Hunt, *Epoch of the Nineteenth Century*, 5 (showing photo of the Hunts' home); P. B. Davis to ?, 10 Jan. 1883, frames 244-45, roll 13, STF Series; Perley B. Davis, "Mrs. Hunt's Call to the Work," *School Physiology Journal* 15 (1906): 149. On middle-class housing and associational patterns during these years, see Stuart Blumin, *The Emergence of the Middle Class: Social Experience in the American City, 1760-1900* (Cambridge, Eng.: Cambridge University Press, 1989), chs. 5-6.

11. Mary A. Livermore to ?, 26 Dec. 1894, frames 66-68, roll 66, STF Series; Hunt diary, 15 Sept. 1879, frame 20; Mary H. Hunt, "Lecture Methods" (MS, n.d. [1883?]), frame 857, both in roll 13, STF Series.

12. Hunt testimony to [Liquor] License Committee, Massachusetts State House, 10 Feb. 1879, frames 685-90, STF Series. On competing visions of law and society within the American prohibition movement, see Richard F. Hamm, *Shaping the Eighteenth Amendment: Temperance Reform, Legal Culture, and the Polity, 1880-1930* (Chapel Hill: University of North Carolina Press, 1995).

13. The same pattern was visible nationwide. In 1875, three states banned the sale of alcohol; by 1890, just three more had joined the fold. Hamm, *Shaping the Eighteenth Amendment*, 25.

14. Mary H. Hunt, "Waiting for a Verdict" (MS, n.d. [1883?]), frame 841, roll 13, STF Series.

15. Julia Colman, "The Work among the Children," in *One Hundred Years of Temperance: A Memorial Volume of the Centennial Temperance Conference* (New York: National Temperance Society and Publishing House, 1886), 260-66; Mitchell, "'The

North and South Here Meet,'" 223; "Temperance in the Schools," *Rochester Daily Union and Advertiser,* 9 April 1883, p. 2.

16. Hunt, "Lecture Methods," frame 565; Transeau biography, frame 464, roll 13, STF Series; Matthew J. Brennan, "A Historical Investigation of Teaching Concerned with the Effects of Alcohol upon the Human Body," (Ph.D. diss., Teachers College, Columbia University, 1949), 7–34, 37–39; James R. Turner, "The American Prohibition Movement, 1865–1897" (Ph.D. diss., University of Wisconsin, 1972), 70–75.

17. Hunt address, n.t., (MS, 1879), frames 747–83, roll 13, STF Series; National Woman's Christian Temperance Union, *Plan of Work for the Introduction of Scientific Temperance Text Books into Colleges, Normal, and Public Schools* (n.p., 1879), frames 1–15, roll 8, STF Series; *Report of the Committee of the Woman's National Christian Temperance Union on Temperance Text-Books in Schools and Colleges* (n.p., 1880), frames 18–20, roll 8, STF Series.

18. *Minutes of the Eighth Convention of the National Woman's Christian Temperance Union* (Brooklyn: Union-Argus, 1881), 41; *Tenth Annual Report of the W.C.T.U. of Pennsylvania* (Pittsburgh: Stevenson and Foster, 1884), 97; Address of Mrs. M. L. Lathrop, quoted in "M.B.W.," "Tenth Annual Convention of the Woman's Christian Temperance Union of Michigan, Held at Adrian, May 23, 24, 25, 26, 27," *Union Signal,* 7 June 1883; clipping from *Christian at Work,* 18 Oct. 1883, frame 143, roll 8, STF Series; Editorial (n.t.), *Michigan School Moderator* 4 (14 Feb. 1884); Mary Hunt, "Difficulties to be Overcome in Securing the Revision of Temperance Text Books to WCTU Standards," *Union Signal,* 22 Sept. 1887.

19. Mary H. Hunt, *A History of the First Decade of the Department of Scientific Temperance Instruction . . . ,* 3rd ed. (Boston: George E. Crosby, 1892), v.

20. Hunt diary, 3 Feb. 1881, frame 102, roll 13, STF Series; Alice M. Guernsey, "New England Letter," *Union Signal,* 25 Feb. 1886; J. C. Perkins, *History of Scientific Temperance Instruction in Vermont* (n.p., n.d.), frame 646, roll 14, STF Series; Mary H. Hunt, *A History of the First Decade of the Department of Scientific Temperance Instruction, . . .* 2nd ed. (Boston: George E. Crosby, 1891), 9. On the Vermont law, see also *Vermont House Journal,* 2 Nov. 1882, pp. 163–64; ibid., 4 Nov. 1882, pp. 171–72; Mary H. Hunt, "Temperance in the Public Schools," in *One Hundred Years of Temperance,* 254–55.

21. Hale, "A Bit of Modern Chivalry," frame 305; Hunt, "Temperance in the Public Schools," 254.

22. *Union Signal,* 1 Feb. 1883; *Minutes of the Tenth Convention of the National Woman's Christian Temperance Union* (Cleveland: Home Publishing, 1883), lxxxiv; Transeau biography, frame 102, roll 14, STF Series; *Union Signal,* 3 March 1883.

23. For descriptions of Hunt's "Michigan"-type campaigns in other states, see, e.g., *Minutes of the Tenth Convention of the National WCTU,* lxxx [New Hampshire]; clipping, *Our Work,* March 1884, frame 170, roll 8, STF Series [Brooklyn]; Transeau biography, frame 108, roll 14, STF Series [Pennsylvania]; Mrs. C. S. Winchell, "Scientific Temperance Instruction in Schools and Colleges," *Report of the Tenth Annual Convention of the W.C.T.U. of the State of Minnesota* (Red Wing: Red Wing Printing Co., 1886), 19–20.

24. Mary F. Lovell, "The Work of a Remarkable Woman and Its Results" (MS, April 1927), frames 168–69, roll 66, STF Series; "Temperance Instruction," *Washington Post,* 21 April 1886, p. 1; Hunt, *Epoch of the Nineteenth Century,* 13–18; *Minutes of the Tenth Convention of the National WCTU,* lxxiii. For other examples of campaigns that failed for lack of constituent work, see Mrs. C. S. Winchell, "Report of the State Superintendent of S.T.I.," *Report of the Eleventh Annual Convention of the Woman's Christian Temperance Union of the State of Minnesota* (Red Wing: Red Wing

Printing Co., 1887), 75-76; clipping, *The Organizer*, 1 June 1883, frame 124, roll 8, STF Series.

25. Ballard L. Campbell, *Representative Democracy: Public Policy and Midwestern Legislatures in the Late Nineteenth Century* (Cambridge: Harvard University Press, 1980), 105-14; "Pilot Fees at New York," *New York Times*, 30 Jan. 1884, p. 1; "The General Assembly," *Iowa State Register* [Des Moines], 6 Feb. 1886, p. 8; Weeks, "Report on Scientific Temperance," 66; *Legislative Record* [Penn.] 1 (31 March 1885): 655. For other examples of anti-STI satires, see *St. Louis Post-Dispatch*, 19 Feb. 1885, p. 4; Edison Putman, "The Prohibition Movement in Idaho, 1863-1934" (Ph.D. diss., University of Idaho, 1979), 81.

26. "Consistency in Legislation," *New Haven Register*, 25 Feb. 1886, p. 2; *Legislative Record* [Penn.] 1 (31 March 1885): 662; *Journal of the Assembly of the State of New York, at Their One Hundred Nineteenth Session* (1896), 485, 728; *Journal of the House of Representatives of the Fortieth General Assembly of the State of Illinois* (1897), 1013; *Journal of the Senate of the Fortieth General Assembly of the State of Illinois* (1897), 828; N. L. Caminade, "New Jersey Redeemed," *Union Signal*, 10 May 1894; Hunt, *History of the First Decade*, 2nd ed., 9. In most states, STI bills passed by such overwhelming margins that newspapers and legislative journals did not even report roll calls on them. In the instances where roll calls were recorded, as shown below, STI generally united the Republicans and split the Democrats:

State/Yr/H-S	RF	RAg	RAb	DF	DAg	DAb
PA/1885/House	106	11	24	25	27	8
IL/1885/Senate	21	0	?	14	4	?
NB/1885/House	51	21	6	3	15	3
NB/1885/Senate	20	4	1	2	5	1
CT/1886/Senate	12	1	0	5	6	0
IA/1886/Senate	27	0	?	2	7	?
IL/1887/Senate	21	0	9	5	0	16
CO/1887/Senate	22	4	?	8	14	?
NY/1896/House	89	0	14	14	22	10
IL/1897/House	56	12	?	20	17	?
IL/1897/Senate	34	0	5	4	1	6
OH/1900/House	53	0	12	25	2	18
OH/1900/Senate	14	0	6	5	0	6

Key: RF = Republicans For; RAg = Republicans Against; RAb = Republicans Abstaining; DF = Democrats For; DAg = Democrats Against; DAb = Democrats Abstaining

Sources: *Legislative Record* [Penn.] (31 March 1885): 662; *Journal of the Senate of the Thirty-Fourth General Assembly of the State of Illinois* (1885), 262-63; *House Journal of the Legislature of the State of Nebraska. Nineteenth Regular Session* (1885), 1223-24; *Senate Journal of the Legislature of the State of Nebraska. Nineteenth Regular Session* (1885), 391; *Journal of the Senate of the State of Connecticut* (1886), 573; "The

158   DISTILLING DEMOCRACY

General Assembly," *Iowa State Register* [Des Moines], 6 Feb. 1886, p. 8; *Journal of the Senate of the Thirty-Fifth General Assembly of the State of Illinois* (1888), 436, 403; William E. West, "Dry Crusade: The Prohibition Movement in Colorado, 1858–1933" (Ph.D. diss., University of Colorado, 1971), 120; *Journal of the Assembly of the State of New York, at Their One Hundred Nineteenth Session* (1896), 485, 728; *Journal of the House of Representatives of the Fortieth General Assembly of the State of Illinois* (1897), 1013; *Journal of the Senate of the Fortieth General Assembly of the State of Illinois* (1897), 828; *Journal of the House of Representatives of the State of Ohio for the Regular Session of the Seventy-Fourth General Assembly* (1900), 935; *Journal of the Senate of the State of Ohio for the Regular Session of the Seventy-Fourth General Assembly* (1900), 672.

27. "Under the Power of the Saloon," *Iowa State Register* [Des Moines], 7 Feb. 1886, p. 4; Robert Bader, *Prohibition in Kansas: A History* (Lawrence: University Press of Kansas, 1986), 73, 99; *Union Signal*, 10 April 1884.

28. *Thirteenth Annual Report of the Woman's Christian Temperance Union of Michigan* (Grand Rapids: I. S. Dygert, 1886), 11–13; *Fiftieth Annual Report of the Superintendent of Public Instruction of the State of Michigan* (Lansing: Thorp and Godfrey, 1887), iii–iv.

29. Carl Kaestle, *Pillars of the Republic: Common Schools and American Society, 1780-1860* (New York: Hill and Wang, 1983), chs. 6-7; David Tyack, et al., *Law and the Shaping of Public Education, 1785-1954* (Madison: University of Wisconsin Press, 1987), ch. 2, esp. 57-59; David Tyack and Elisabeth Hansot, *Managers of Virtue: Public School Leadership in America, 1820-1980* (New York: Basic, 1982), pt. 1; Wayne E. Fuller, *The Old Country School: The Story of Rural Education in the Middle West* (Chicago: University of Chicago Press, 1982), chs. 5-8.

30. In 1884, for example, a "special temperance committee" of the National Education Association recommended "the hearty cooperation" of the NEA "in making [STI] legislation general throughout the land." Scientific Temperance also won special praise from NEA president Thomas Bicknell, who invited WCTU leader Frances Willard to address the association's national convention. *Journal of Proceedings and Addresses of the National Education Association* 23 (1884): 15; *Education* 5 (1885): 311-13; Frances Willard, "Lead Pencil Letter—No. III," *Union Signal*, 31 July 1884. See also Chapter Four of this book.

31. *Biennial Report of the Superintendent of Public Instruction of the State of Iowa* [1886-87] (Des Moines: George E. Roberts, 1887), 161; *Report of the Commissioner of Education for the Year 1887-88* (Washington, D.C.: GPO, 1889), 101; Minutes of the Ann Arbor [MI] WCTU, 3 Feb. 1885, Box 1, Woman's Christian Temperance Union of Michigan Records, Bentley Historical Library, University of Michigan, Ann Arbor, Michigan [hereafter Michigan WCTU]; *Report of the Commissioner of Education for the Year 1889-90*, 2 vols. (Washington, D.C.: GPO, 1891), 1:389-95; *First Biennial Report of the Superintendent of Public Instruction, to the Governor of North Dakota* [1889-90] (Bismarck: Tribune, 1890), 108; *Forty-Fourth Annual Report of the Board of Education of the City of New York* [1885] (New York: Hall of the Board of Education, 1886), 25; Rochester Public Schools, *Thirty-Seventh Annual Report of the Board of Education* [1883-84] (Rochester: Sunday Herald Job Office, 1884), 110; *Thirty-Seventh Annual Report of the Board of Education of the City of Syracuse* [1884-85] (Syracuse: Standard Book and Job Print, 1884).

32. In Minnesota, for example, all but three of the twenty-four representatives who were born in Ireland, Germany, and Norway cast their ballots in favor of STI. *Journal of the House of the 25th Session of the Legislature of the State of Minnesota* (1887),

426; Victor Hjortsberg, *Legislative Manual for the State of Minnesota* (St. Paul: J. W. Cunningham, 1887), 538-40.

33. "Practical Working of the Ainsworth School Physiology Law in New York State," *School Physiology Journal* 6 (Oct. 1896): 21; "The Schoolbooks Question," *Abendpost*, 8 March 1897, file IA1a, box 10, Chicago Foreign Language Press Survey, Joseph Regenstein Library, University of Chicago, Chicago, Illinois; *Hearings before the Subcommittee on the Judiciary. United States Senate Sixty-Fifth Congress, Second Session on S. 3529* . . . (Washington, D.C.: GPO, 1918), 593-94, 674.

34. *Twenty-First Annual Report of the State Board of Education Showing the Condition of the Public Schools* (Annapolis: James Young, 1888), 144; A. H. Plumb to H. P. Bowditch, 20 Jan. 1896, frame 54, roll 11, STF Series; Catherine Gilbert Murdock, "Domesticating Drink: Women and Alcohol in Prohibition America, 1870-1940" (Ph.D. diss., University of Pennsylvania, 1995), pt. 2; Minutes of the Ann Arbor [MI] WCTU, 3 Feb. 1885, Box 1, Michigan WCTU; Sarah S. Low to MHH, 12 May 1902, frames 521-22, roll 12, STF Series; *Michigan School Moderator* 4 (21 Feb. 1884): 45; *Report of the Superintendent of Public Instruction of Washington Territory* (Olympia: Thomas H. Cavanaugh, 1887), 64; *Biennial Report of Iowa* (1886-87), 32.

35. Hannah A. Foster, "Ohio: Persistent, Progressive Efforts," *Union Signal*, 25 June 1891; T. P. Bagby, "Scientific Temperance in Virginia," *Union Signal*, 2 May 1901; Mary E. Newton, "Scientific Temperance Instruction," *Union Signal*, 16 April 1891; Anna M. Greene, "Pennsylvania: State Notes and Scientific Institutes," *Union Signal*, 28 May 1891; *Sunday News* [Detroit], 3 May 1891, frame 360, roll 18, STF Series; "Difficulties, and How to Overcome Them," *Union Signal*, 25 June 1891.

36. Mary H. Hunt, "Scientific Temperance Instruction," *Union Signal*, 9 Jan. 1890; Woman's Christian Temperance Union, *The Pathfinder or National Plans for Securing Scientific Temperance Education in Schools and Colleges* (New York: A. S. Barnes, 1885), 19-21; Hunt, *History of the First Decade*, 2nd ed., 58, 53.

37. Hunt, *History of the First Decade*, 2nd ed., 41-42; Mary H. Hunt, "Report of the National Department of Scientific Temperance Instruction," *Union Signal*, 14 Nov. 1895; Mary Hunt, "Law Without a Penalty Is Only Advice," *Union Signal*, 14 Jan. 1886.

38. *Colorado School Journal* 2 (April 1887): 5; ibid. 2 (June 1887): 2; Jason E. Hammond to Caroline Humphrey, 10 May 1898, box 1, Scientific Temperance Federation Papers, New York Public Library, New York, New York [hereafter STF Papers]; *Biennial Report of Iowa* (1886-87), 16-18; *Fifty-First Annual Report of the Superintendent of Public Instruction of the State of Michigan* [1887] (Lansing: Thorp and Godfrey, 1888), x-xi; *Fifty-Fourth Annual Report of the Superintendent of Public Instruction of the State of Michigan* [1890] (Lansing: Robert Smith and Co., 1891), lv-lvi. In STI's first two decades, I found only two instances—both in Pennsylvania—of school boards losing money for failing to comply with it. Anna M. Greene, "Pennsylvania: State Notes and Scientific Institutes," *Union Signal*, 28 May 1891.

39. Hunt, *History of the First Decade*, 2nd ed., 27; *Union Signal*, 15 April 1886; MHH to Dr. L. Anna Ballard, 17 Jan. 1889, box 1, STF Papers; *Union Signal*, 22 April 1886; Clara B. Hoffman and Lida B. Ingalls, "The St. Louis Health Primer," *Union Signal*, 20 Jan. 1887. See also "History of the Indorsement and Advisory Board" (MS, n.d.), frame 153, roll 15, STF Series.

40. "A Word For Teachers," *Our Message*, n.d. [1888?], frame 821, roll 14; Edward F. Adams to L. S. Tobey, 7 April 1886, frame 91, roll 15; Adams to Tobey, 3 Aug. 1886, frame 95, roll 15; Adams to Tobey, 27 Jan. 1887, frames 630-33, roll 16, all in STF Series.

41. A. A. Meader to MHH, 14 July 1885, frame 93, roll 15; Mary P. Buehler to MHH, 26 June 1885, frame 89, roll 15; Meader to MHH, 2 June 1886, frame 87, roll 15; MHH to Mary Lovell, 9 Nov. 1885, frames 409-10, roll 17; Lovell to Philadelphia Board of Education, 29 Nov. 1886, frame 101? (unintell.), roll 15, all in STF Series. On competition and corruption in the textbook industry, see John Tebbel, *A History of Book Publishing in the United States*, 2 vols. (New York: R. K. Bowker, 1975), 2: 559-86; Fuller, *Old Country School*, 93-96; Tyack, et al., *Law and the Shaping of Public Education*, 99-105.

42. Mrs. Joseph D. Weeks to MHH, 21 April 1886, frames 407-14, roll 3; Frances Swift to MHH, 7 Oct. 1885, frame 103, roll 15; George R. Cathcart to A. S. Barnes, 13 Nov. 1885, frames 269-70, roll 3, all in STF Series.

43. J. Ellen Foster to MHH, 8 Dec. 1885, frames 754-63, roll 2, STF Series.

44. MHH to J. F. Weeks, 4 Jan. 1890, frame 544, roll 16, STF Series; *Minutes of the Seventh Convention of the National Woman's Christian Temperance Union* (New York: National Temperance Society and Publication House, 1880), 96-97; *Minutes of the Eleventh Convention of the National Woman's Christian Temperance Union* (Chicago: Woman's Temperance Publishing Association, 1884), 62; Hunt, *Epoch of the Nineteenth Century*, 7. For more detailed descriptions of the WCTU's organizational structure, see Jonathan Zimmerman, "'The Queen of the Lobby': Mary H. Hunt, Scientific Temperance, and the Dilemma of Democratic Education in America, 1879-1924," (Ph.D. diss., Johns Hopkins University, 1993), 378-80; K. Austin Kerr, *Organized for Prohibition: A New History of the Anti-Saloon League* (New Haven: Yale University Press, 1985), 48-50.

45. Letter by Mary Leavitt, *Union Signal*, 12 July 1883; Emily M. Laughlin to MHH, 18 Sept. 1883, frames 792-93, roll 2, STF Series. See also Laughlin to MHH, 3 Oct. 1883, frames 798-99, roll 2, STF Series.

46. C. C. Alford to MHH, 24 Oct. 1883, frames 259-64; Mary Burt to C. C. Alford, 23 Oct. 1883, frames 265-66, both in roll 3, STF Series; J. Ellen Foster, *Scientific Temperance Instruction in the Public Schools* (Boston: Frank Wood, 1886), 11-12; *Our Work*, March 1884, frame 170, roll 8, STF Series; Hunt, *Epoch of the Nineteenth Century*, 9.

47. Katharine L. Stevenson, *A Brief History of the Woman's Christian Temperance Union* (Evanston, Ill.: The Union Signal, 1907), 32-34; "How the Temperance Instruction Laws were secured," (MS, n.d.), frame 118, roll 66, STF Series; "Minutes of the 11th Convention," *Union Signal*, 30 Oct. 1884; clipping, *West Virginia Freeman* [Parkersburg], 11 Nov. 1885, Willard Scrapbook No. 23, National Woman's Christian Temperance Union Scrapbooks, Frances E. Willard Memorial Library, Evanston, Illinois. On the partisan dispute, see Ruth Bordin, *Woman and Temperance: The Quest for Power and Liberty, 1873-1900* (Philadelphia: Temple University Press, 1981), 123-33; Kerr, *Organized for Prohibition*, 54-65.

48. "Minutes of the 12th Convention," *Union Signal*, 19 Nov. 1885; "Fear versus Facts," *Union Signal*, 26 Nov. 1885; Mary Hunt, "Law without a Penalty Is Only Advice," *Union Signal*, 14 Jan. 1886; *Minutes of the Twelfth Convention of the National Woman's Christian Temperance Union* (Brooklyn: Martin and Niper, 1885), 33-34. Led by J. Ellen Foster, a group of antiendorsement members would soon bolt the WCTU to forge a separate "Non-Partisan" organization. But Hunt remained in the mother union, arguing that secession could only weaken Scientific Temperance. "Think of it, two little Unions in a small town spending their force in antagonizing each other," Hunt wrote in 1886. "Think of two local superintendents of each Department struggling to get hold of the same work in the same place." Kerr, *Organized for*

Prohibition, 67-73; Bordin, *Woman and Temperance,* 129-31; MHH to ?, 10 Nov. 1886, frame 731, roll 14, STF Series.

49. A. S. Barnes to MHH, 22 March 1884; MHH to Frances Willard, n.d. [1884?], both at frame 110, roll 15, STF Series; Mattie F. Weeks to MHH, 8 Feb. 1885, frame 361, roll 14, STF Series; ? to MHH, n.d. [1886], box 1, Scientific Temperance Federation Papers, New York Public Library, New York, New York; Mary Lovell to MHH, 19 Aug. 1886, frame 202, roll 15, STF Series. See also review of Smith's *Primer of Physiology and Hygiene* in *Union Signal,* 4 Feb. 1886; M. H. Hunt, "Why I Do Not Endorse Smith's Physiologies," *Union Signal,* 25 March 1886. On Willard and alcohol use, see Ruth Bordin, *Frances Willard: A Biography* (Chapel Hill: University of North Carolina Press, 1986), 68; idem, *Woman and Temperance,* 42-43, 99.

50. Nancy Hardesty, *Women Called to Witness* (Nashville: Abington Press, 1984), 14-15; Bordin, *Frances Willard,* 14-15, 19. The Willards' trek to Ohio coincided with the rise of the Washingtonians, whose noisy temperance meetings echoed the rhythms of Finneyite revivalism. Frances's father was an honorary member of the society, which also promoted cold-water parades and pledge signings for children. Bordin, *Frances Willard,* 68.

51. Willard would later embrace the holiness doctrines of Methodist evangelist Phoebe Palmer, who told her followers they could achieve "entire sanctification" through study, prayer, and service. See Ian Tyrrell, *Woman's World/Woman's Empire: The Woman's Christian Temperance Union in International Perspective, 1880-1930* (Chapel Hill: University of North Carolina Press, 1991), 23-24; Bordin, *Frances Willard,* 156.

52. Bordin, *Woman and Temperance,* ch. 2; Anna A. Gordon, *The Beautiful Life of Frances Willard* (Chicago: Woman's Temperance Publishing Association, 1898), 126; Hardesty, *Women Called to Witness,* p. 21; Rebecca Edwards, "Gender in American Politics, 1880-1900" (Ph.D. diss., University of Virginia, 1995), 102, 130, 98; Frances E. Willard, *Glimpses of Fifty Years: The Autobiography of an American Woman* (Chicago: Woman's Temperance Publishing Association, 1889), 403.

53. Hunt, "Temperance in the Public Schools," 256; "Temperance in the Schools," *Rochester Daily Union and Advertiser,* 9 April 1883; *Mount Vernon* [Iowa] *Hawkeye,* ? Dec. 1881, frame 53, roll 8, STF Series; "Teaching Temperance in the Public Schools" (MS, n.d. [1897?]), folder 3, Catharine Dunham Papers, Arthur and Elizabeth Schlesinger Library, Radcliffe College, Cambridge, Massachusetts; E. A. Rand, "Temperance Instruction in Public Schools," *Pennsylvania School Journal* 33 (Oct. 1884): 142.

54. Stevenson, *Brief History,* 32-34; Frances Willard, *Hints and Helps in Our Temperance Work* (New York: National Temperance Society, 1875), 30-32; Lucia E. F. Kimball, *Why Special Temperance Teaching in the Sunday School?* (Chicago: Woman's Temperance Publishing Association, n.d.), 6.

55. *Minutes of the Eleventh Convention of the National WCTU,* xxv, front leaf; "Informal Gathering" (MS, May 1906), frame 25, roll 39, STF Series; Frances Willard, *Occupations for Women* (New York: Success Co., 1897), 432; Zimmerman, "The Queen of the Lobby," 484-85; Judith B. Erikson, "Making King Alcohol Tremble: The Juvenile Work of the Woman's Christian Temperance Union, 1874-1900," *Journal of Drug Education* 18 (1988): 338-40; Frances Willard, *Woman and Temperance* (Hartford: Park Publishing, 1883), 607-8; letter from Anna Gordon, *Union Signal,* 13 Nov. 1884.

56. Erikson, "Making King Alcohol Tremble," 344-47; Helen E. Tyler, *Where Prayer and Purpose Meet* (Evanston, Ill.: Signal Press, 1949), 227-28. Addressing the World's WCTU Convention at the 1893 Columbian Exposition in Chicago, Willard

praised the LTL as "the most progressive method" of temperance education. The Legion obtained over 300,000 signed pledge cards for the Exposition, festooning them across the Children's Building and the main WCTU exhibit. Both displays seem to have ignored Scientific Temperance Instruction, much to Mary Hunt's annoyance. Frances Willard, *Address before the Second Biennial Convention of the World's Woman's Christian Temperance Union* . . . (Chicago: Woman's Temperance Publishing Association, 1893), 10-11; Frances Willard, *Do Everything: A Handbook for the World's White Ribboners* (Chicago: Woman's Temperance Publishing Association, n.d. [1895?]), 61.

57. This strategy prefigured Hunt's efforts to counter enemy scientists, educators, and theologians: "fight fire with fire." Just as Hunt enlisted sympathetic experts to rebut professional attacks, so would she recruit grass-roots allies to fend off localist ones. See Chapters Three through Six of this book.

58. Frances Whitaker, "A History of the Ohio Woman's Christian Temperance Union, 1874-1920" (Ph.D. diss., Ohio State University, 1971), 315-18; *Scientific Temperance Bulletin* (n.p. [Ohio], 1889), 33; Sarah Pratt to MHH, 24 May 1892, frames 285-88, roll 3, STF Series; Pratt to MHH, 24 Aug. 1892, frames 285-88, roll 3, STF Series. The *Bulletin* mirrored Hunt's *School Physiology Journal* in many ways, stressing the physiological basis for strict teetotal instruction. It even reprinted Hunt's demand for texts "in the hands of pupils"—a rich irony, given Leiter's earlier capitulation on this point. Yet it lacked the hands-on advice that Hunt's publications had provided: as district superintendent Sarah Pratt complained to Hunt, the *Bulletin* functioned more as a "Scientific magazine" than as a practical guide to monitoring STI. Mattie Weeks to MHH, 9 Dec. 1889, frames 458-61, roll 3, STF Series; *Scientific Temperance Bulletin* (n.p. [Ohio], 1889]), 26-27; Sarah Pratt to MHH, 24 May 1892, frames 285-88, roll 3, STF Series. Located at the Library of Congress, this single issue (month unknown) was the only trace of the *Bulletin* I could find. Yet Pratt's letters to Hunt indicate that Leiter continued to publish it at least through 1892, if not longer. See also E. J. Gray to MHH, 23 Aug. 1892, frames 302-6, roll 3, STF Series.

59. Pratt to MHH, 10 Aug. 1891, frames 280-82, roll 3, STF Series.

60. Pratt to MHH, 2 June 1892, frames 289-90; E. J. Gray to MHH, 23 Aug. 1892, frames 302-6, both in roll 3, STF Series; Whitaker, "History of the Ohio WCTU," 321-23; Pratt to ?, 24 Aug. 1892, frame 307, roll 3, STF Series; Mattie Weeks to MHH, 9 Dec. 1889, roll 16, STF Series; *Minutes and Reports of the Woman's Christian Temperance Union of the State of Illinois, at the Seventeenth Annual Convention* (Chicago: Woman's Temperance Publishing Association, 1890), 137.

61. *Eighteenth Annual Report of the Woman's Christian Temperance Union of Michigan* (Grand Rapids: Western Michigan Printing Co., 1891), 15; Minutes of the 10th Convention, 19-22 September 1893, box 8, Eleventh District Records, Michigan WCTU; "Educational Agencies," *Union Signal*, 20 March 1890; Minutes, 26 April 1887, Woman's Christian Temperance Union of Alfred [N.Y.] Records, Alfred University Library, Alfred, New York [hereafter Alfred WCTU].

62. See, e.g., Hunt, *Epoch of the Nineteenth Century*, 58-59; MHH to Irene G. Adams, 4 Oct. 1901, frames 189-91, roll 7, STF Series.

63. "Educational Agencies," *Union Signal*, 20 March 1890.

64. See, e.g., Minutes, 26 April 1887; 27 March 1888; 23 Jan. 1889; 30 Nov. 1889; 18 Feb. 1890, Alfred WCTU.

65. Clara C. Coombs, "Scientific Temperance Instruction," *Minutes and Reports of the Woman's Christian Temperance Union of the State of Illinois, at the Eighteenth Annual Convention* (Chicago: Woman's Temperance Publishing Association, 1891), 125; Minutes, 27 May 1897, Orleans [VT] Records, Woman's Christian Temper-

ance Union of Vermont Records, Bailey-Howe Library, University of Vermont, Burlington, Vermont.

66. Program for "Scientific Temperance Instruction Evening," 1 Nov. 1897, frame 580; Frances Willard to MHH, 6 Sept. 1897, frame 502, both in roll 1, STF Series; Tyler, *Where Prayer and Purpose Meet,* 230–31; *Report of the National Woman's Christian Temperance Union. Twenty-Fourth Meeting* (Chicago: Woman's Temperance Publishing Association, 1897), 62. For Hunt's reactions to the reduction of her appropriation, see MHH to "Member of the Executive Committee," frames 426–34, roll 17, STF Series; MHH to Rebecca Chambers, 6 Jan. 1898, frame 629, roll 13, STF Series; MHH to Cornelia B. Forbes, 14 Feb. 1898, frame 736, roll 15, STF Series.

3. "WHEN THE DOCTORS DISAGREE"

1. Merrill C. Ward to Mary Hunt, 4 April 1901, frames 864–65, roll 2, Temperance and Prohibition Papers, Ohio Historical Society (joint Ohio Historical Society—Michigan Historical Collections), Scientific Temperance Federation Series [hereafter STF Series].

2. Merrill C. Ward to Wilbur O. Atwater, 15 April 1901, box 5, Wilbur O. Atwater Papers, Olin Library, Wesleyan University [hereafter Atwater Papers].

3. On competing expert claims in modern America, see especially Brian Balogh, *Chain Reaction: Expert Debate and Public Participation in the Development of American Commercial Nuclear Power* (Cambridge, Eng.: Cambridge University Press, 1991).

4. Mary O. Furner, *Advocacy and Objectivity: A Crisis in the Professionalization of American Social Science, 1865-1905* (Lexington: University Press of Kentucky, 1975); Thomas L. Haskell, *The Emergence of Professional Social Science: The American Social Science Association and the Nineteenth-Century Crisis of Authority* (Urbana: University of Illinois Press, 1977).

5. This is the theme of a vast literature in medical history, most prominently *The Therapeutic Revolution: Essays in the Social History of American Medicine,* ed. Morris J. Vogel and Charles E. Rosenberg (Philadelphia: University of Pennsylvania Press, 1979); Paul Starr, *The Social Transformation of American Medicine* (New York: Basic Books, 1982).

6. Philip J. Pauly, "The Struggle for Ignorance about Alcohol: American Physiologists, Wilbur Olin Atwater, and the Woman's Christian Temperance Union," *Bulletin of the History of Medicine* 64 (1990): 366–92, esp. 390.

7. David Hollinger, "Inquiry and Uplift: Late Nineteenth-Century American Academics and the Moral Efficacy of Scientific Practice," in *The Authority of Experts: Studies in History and Theory,* ed. Thomas L. Haskell (Bloomington: Indiana University Press, 1984), 142–56; Pauly, "Struggle for Ignorance," esp. 368.

8. Mrs. J. Frank Smith, *Co-Worker's Letter* (n.p., 1901), frame 242, roll 16, STF Series.

9. Warner, "Science in Medicine," 52.

10. Mary Hunt, et al., *Science the Arbiter: A Reply to Certain Criticisms of Dr. L. J. Lautenbach and His Committee on School Textbooks in Physiology and Hygiene in the Pennsylvania Medical Journal* (n.p.: Pennsylvania Woman's Christian Temperance Union, 1904).

11. Thomas Bender, "Science and the Culture of American Communities: The Nineteenth Century," *History of Education Quarterly* 16 (1976): 70; Daniel Kevles, "The Physics, Mathematics, and Chemistry Communities: A Comparative Analysis," in *The Organization of Knowledge in Modern America, 1860-1920,* ed. Alexandra Oleson and John Voss (Baltimore: Johns Hopkins University Press, 1979), 152; Thomas

Haskell, "Power to the Experts," *New York Review of Books,* 17 Oct. 1977, p. 33. Haskell's essay examines social science, but his point is essentially the same one that historians of the natural sciences make: greater complexity of knowledge made older forms of explanation obsolete.

12. On "elite science," see esp. Daniel J. Kevles, *The Physicists: The History of a Scientific Community in Modern America* (New York: Knopf, 1978).

13. *Report of the Commissioner of Education for the Year 1886-87* (Washington: GPO, 1888), 217; *Minutes of the Twelfth Convention of the National Woman's Christian Temperance Union* (Brooklyn: Martin and Niper, 1885), appendix, p. 9; Erwin F. Smith, "Physiology Text Books, Approved by State Board of Health," *Michigan School Moderator* 5 (18 June 1885): 775; "Text-Books in Physiology," *Wisconsin Journal of Education* 15 (Aug. 1885): 338.

14. "Scientific Temperance," *Chautauquan* 9 (Dec. 1888): 162-65. See also "Physicians' Petition for Scientific Temperance Instruction in Public Schools" (MS, n.d. [1885]), frame 489, roll 9, STF Series.

15. "Total-Abstinence Teaching in the Schools," *Science* 7 (5 Feb. 1886): 115-16; *Popular Science Monthly* 25 (Aug. 1884): 553-56. See also Pauly, "Struggle for Ignorance," 373-74.

16. *The Liquor Problem: A Summary of Investigations Conducted by the Committee of Fifty, 1893-1903,* ed. John S. Billings, et al. (Boston and New York: Houghton, Mifflin, 1905), 3-4. See also *Physiological Aspects of the Liquor Problem,* 2 vols., ed. John S. Billings (Boston and New York: Houghton, Mifflin, 1903); Pauly, "Struggle for Ignorance," 373-78.

17. As this sample indicates, the APS defined "physiology" broadly enough to include virtually any type of laboratory research and training. See Robert Frank, "Americans in German Laboratories, 1865-1914," in *Physiology in the American Context, 1850-1940,* ed. Gerald L. Geison (Bethesda, Maryland: American Physiological Society, 1987), 12-13. On American physiology in general, see also W. Bruce Fye, *The Development of American Physiology: Scientific Medicine in the Nineteenth Century* (Baltimore: Johns Hopkins University Press, 1987).

18. On the importation of the German research ethic, see Thomas N. Bonner, *American Doctors and German Universities: A Chapter in International Intellectual Relations, 1870-1914* (Lincoln: University of Nebraska Press, 1963); Frank, "Americans in German Laboratories."

19. Pauly, "Struggle for Ignorance," 380-82; Billings, *Physiological Aspects,* 1: xvii. See also W. O. Atwater, "The Nutritive Value of Alcohol," ibid., 2: 169-348.

20. W. O. Atwater [hereafter WOA] to J. S. Billings, 1 Nov. 1897, box 8, Atwater Papers.

21. Mary H. Hunt [hereafter MHH] to Mary F. Lovell, 20 Feb. 1892, frame 319, roll 9, STF Series; William A. Mowry to MHH, 13 Nov. 1903, box 1, Scientific Temperance Federation Papers, New York Public Library, New York, New York [hereafter STF Papers].

22. *School Physiology Journal* 6 (March 1897), "Extra Edition." Other authors of pro-STI statements included Boston clinician Henry O. Marcy, former president of both the American Medical Association and the American Academy of Medicine; E. E. Montgomery, president of the Pennsylvania Medical Society and professor at Jefferson Medical College; and George F. Shrady, editor of the *Medical Record* in New York.

23. *Report of the National Woman's Christian Temperance Union. Twenty-Seventh Meeting* (Chicago: Woman's Temperance Publishing Association, 1900), 214; Pauly, "Struggle for Ignorance," 384-85; L. D. Mason to MHH, 4 June 1903, box 1, STF Papers.

24. C. F. Chandler to WOA, 26 April 1900, box 5, Atwater Papers.

25. Henry F. Hewes, "The School Books Defended," *Outlook* 62 (29 July 1899): 706. Hewes republished this defense in a letter to the *Times* dated 5 July 1899 and reprinted in *An Appeal to Truth: An Analysis of Professor Atwater's Statements Regarding the Nutritive Value of Alcohol and of His Tables in "Bulletin 69"* (Boston and New York: n.p., 1900), 13.

26. Atwater based this claim upon the body's alleged maintenance of "nitrogenous equilibrium" when fed small amounts of alcohol; yet Herter and Hall calculated that the alcohol subject in at least one Atwater experiment had actually lost nitrogen, indicating that protein had been squandered rather than saved. Hunt seems to have discerned the same discrepancy on her own in 1899. See Herter's remarks in *Appeal to Truth,* 16; W. S. Hall, "Is Alcohol a Food? (A Reply to Professor W. O. Atwater)," *Quarterly Journal of Inebriety* 22 (1900): 312-21; Pauly, "Struggle for Ignorance," 387; MHH to W. S. Hall, 22 Nov. 1899, frames 832-33, roll 6; MHH to W. S. Hall, 23 Dec. 1899, frames 328-29, roll 17, both in STF Series.

27. Winfield S. Hall, *A Laboratory Guide in Physiology* (Chicago: W. T. Keener and Co., 1896); Winfield S. Hall, *Elementary Anatomy, Physiology, and Hygiene for Higher Grammar Grades* (New York: American Book Co., 1900); Henry F. Hewes, *Anatomy, Physiology, and Hygiene for High Schools* (New York: American Book Co., 1900).

28. Clifton Hodge to WOA, 28 Jan. 1900, box 5; WOA to William H. Welch, 7 March 1900, box 8; Hodge to WOA, 27 Jan. 1900, box 5, all in Atwater Papers.

29. Welch to Herter, 19 March 1900, folder 13, box 27, William H. Welch Papers, The Alan Mason Chesney Archives of the Johns Hopkins Medical Institutions, Baltimore, Maryland [hereafter Welch Papers].

30. Atwater initially asked Welch to seek a retraction from Herter, Welch's former disciple at Johns Hopkins. The move backfired. Welch admitted that he felt "some interest" in defending Atwater's work, "as it is done under the auspices of the Committee of Fifty." Upon reviewing the evidence, however, Welch agreed that Atwater's numbers contradicted the claim of protein protection. Atwater eventually pruned the protein claim from both his laboratory bulletin and from the final subcommittee report, which is explicitly agnostic on the question. Pauly, "Struggle for Ignorance," 387; Welch to Herter, 12 March 1900, 19 March 1900, 29 March 1900, all in folder 13, box 27, Welch Papers; Atwater, "Nutritive Value of Alcohol," 215.

31. "Conversation of Professor W. O. Atwater with Reverend Lyman Abbott, D.D., March 24, 1898 . . ." (transcript); WOA to Bowditch, 25 Oct. 1897, both in box 10, Atwater Papers.

32. Thomas Bonner, "Dr. Nathan S. Davis and the Growth of Chicago Medicine, 1850-1900," *Bulletin of the History of Medicine* 26 (1952): 360-74, esp. 369.

33. Susan Lederer, "Human Experimentation and Anti-Vivisection in Turn-of-the-Century America" (Ph.D. diss., University of Wisconsin, 1987), 7. The ensuing discussion relies upon this dissertation and upon Susan Lederer, "The Controversy over Animal Experimentation in America, 1880-1914," in *Vivisection in Historical Perspective,* ed. Nicolaas Rupke (London: Croom Helm, 1987), 236-58; Patricia Gossell, "William Henry Welch and the Antivivisection Legislation in the District of Columbia, 1896-1900," *Journal of the History of Medicine and Allied Sciences* 40 (1987): 397-419; and James Turner, *Reckoning with the Beast: Animals, Pain, and Humanity in the Victorian Mind* (Baltimore: Johns Hopkins University Press, 1980).

34. For example, Hunt praised the controversial alcohol experiments that Hodge performed upon dogs and cats. The same studies drew angry denunciations from Mary F. Lovell, Hunt's longtime STI assistant but also an activist in the WCTU's antivivisectionist "Band of Mercy." Lovell's dual penchant for STI and antivivisection was an anomaly within the WCTU, where the two departments rarely intersected.

Indeed, there seems to have been a good deal of hostility between them. In one of the rare instances where another WCTU branch infringed upon Hunt's turf, anti-vivisectionists forced through a national resolution urging "that no vivisection experiments be introduced in any school book that we are asked to indorse." To further allay their fears, meanwhile, Hunt was often forced to issue denials that classroom STI demonstrations involved vivisection. Yet her "endorsed" texts continued to commend dissection of frogs and sometimes even of rabbits. "Voices of Authority," *Union Signal*, 3 Oct. 1901; letter from Mary F. Lovell, *Journal of Zoophily* 7 (1898): 28; "Report of the 14th Annual Convention," *Union Signal*, 1 Dec. 1887; Anna E. McGovern, "Effects of Alcohol on the Nervous System," *Iowa Normal Monthly* 14 (1891): 286. For more on Lovell, see Lederer, "The Controversy over Animal Experimentation," 238–39; for examples of animal experiments in the STI classroom, see Chapter Five of this book.

35. Gossell, "William Henry Welch"; Bowditch to WOA, 23 Feb. 1900, box 5, Atwater Papers.

36. Nathan S. Davis, "The Obligations of Medicine to Physiology, and of Physiology to Medicine," *Boston Medical and Surgical Journal* 126 (7 April 1892): 345–47. Bernard's student, John Dalton, introduced vivisection into American physiology. Fye, *Development of American Physiology*, ch. 1.

37. See, e.g., Hall, *Laboratory Guide in Physiology*. See also Winfield S. Hall, *The Method of Teaching Physiology in the Northwestern University Medical School* (Chicago: n.p., 1901).

38. The American Medical Association and the American Academy of Medicine waffled on total-abstinence STI but eventually granted it qualified support. "Shall Alcohol Be Recognized as a Food?" *Journal of the American Medical Association* 34 (31 March 1900): 819; "Report of the Committee on the Teaching of Hygiene in Public Schools," *Bulletin of the American Academy of Medicine* 7 (1905): 29–32. From the start, however, both societies unequivocally condemned government regulation of vivisection. So did state medical societies that supported teetotal teaching, including those in Pennsylvania and Massachusetts. William H. Welch, *Defense of Vivisection* (n.p, n.d. [1896?]), box 126, Welch Papers. See also Gossell, "William Henry Welch."

39. A handful of states did pass laws regulating vivisection in public schools. But most state antivivisection measures failed, causing activists to concentrate upon federal measures. Lederer, "Human Experimentation and Anti-Vivisection," 71, 79. See also Gossell, "William Henry Welch."

40. See, e.g., William T. Sedgwick, "The Modern Subjection of Science and Education to Propaganda," *Science* 15 (1902): 44–54.

41. John Harley Warner, "Ideals of Science and Their Discontents in Late Nineteenth-Century American Medicine," *Isis* 82 (1991): 454–78, esp. 464–68.

42. See, e.g., "Leading Medical Journals of the United States on the Alcohol-A-Food Theory, Since Jan. 1, 1900," *School Physiology Journal* 9 (May 1900), extra ed., p. 4.

43. Hunt address to Wisconsin State Convention (MS, May 1901), frame 139, roll 14, STF Series; W. F. Bynum, "Alcoholism and Degeneration in 19th Century European Medicine and Psychiatry," *British Journal of Addiction* 79 (1984): 68–69; E. Destree, "The Influence of Alcohol on Muscular Work," *Quarterly Journal of Inebriety* 23 (1901): 23; T. D. Crothers to MHH, 23 Oct. 1902, box 1, STF Papers; Jules Morel, "Prevention of Mental Diseases," *American Journal of Sociology* 5 (1899): 81. As historian Barbara Sicherman reminds us, "*psychiatrists,* as they are now defined, did not exist in the late nineteenth century." For purposes of convenience, I use the term to cover neurologists and asylum superintendents, the two groups that Sicherman

subsumes under the designation of "mental hygienists." Barbara Sicherman, "The Paradox of Prudence: Mental Health in the Gilded Age," in *Madhouses, Mad-doctors, and Madmen: The Social History of Psychiatry in the Victorian Era,* ed. Andrew Scull (Philadelphia: University of Pennsylvania Press, 1981), 234n.

44. Arthur B. Patten to WOA, 11 Nov. 1899, box 4, Atwater Papers; W. O. Atwater, "Memorandum to Accompany Dr. Billings' Suggestions for Conclusions by the Sub-Committee" (MS, n.d.), "Committee of Fifty Reports" folder, box 52, John S. Billings Papers, New York Public Library, New York, New York; WOA to Arthur B. Patten, 15 Nov. 1899, box 4, Atwater Papers.

45. John J. Abel, "A Critical Review of the Pharmacological Action of Ethyl Alcohol . . . ," in Billings, *Physiological Aspects,* 2:1-167.

46. See, e.g., George E. Partridge, "Studies in the Psychology of Alcohol," *American Journal of Psychology* 11 (1900): 318-76; John J. Abel to Burt Wilder, 29 Nov. 1903, folder 5, box 2, Burt Green Wilder Papers, Department of Manuscripts and University Archives, Cornell University Library, Ithaca, New York.

47. Destree, "Influence of Alcohol"; August Forel, "The Effect of Alcoholic Intoxication upon the Human Brain and Its Relation to Theories of Heredity and Evolution," *Quarterly Journal of Inebriety* 15 (July 1893): 204; MHH to Edward N. Packard, 13 Sept. 1899, frame 83, roll 6, STF Series. See also D. Boeck, "The Influence of Alcoholic Liquors on Mental Work," *Quarterly Journal of Inebriety* 23 (1900): 48-61; Mary H. Hunt, "Alcohol as a Nerve Tonic," *New Voice,* 27 March 1902, frame 481, roll 17, STF Series.

48. L. D. Mason to MHH, 27 June 1903, box 1, STF Papers; George W. Webster to MHH, 12 Aug. 1902, box 1, STF Papers; G. Stanley Hall to MHH, 29 June 1901, frame 887, roll 2, STF Series; Welch to Herter, 19 March 1900, folder 13, box 27, Welch Papers. For frustrated attempts by Hunt to locate an American scientist who studied alcohol's mental effects, see MHH to H. H. Cushing, 7 Nov. 1901, frames 624-25, roll 7, STF Series. See also MHH to G. M. Gould, 20 Sept. 1901, frame 5; MHH to Harold H. Knett, 20 Sept. 1901, frame 13, both in roll 7, STF Series.

49. "This school is peculiarly American, in history and attainments," exulted Connecticut asylum superintendent T. D. Crothers, probably the best-known inebriest of his era. T. D. Crothers, "New School of Psychology," *Quarterly Journal of Inebriety* 19 (1897): 209-10. On inebriety and inebriests in general, see Jim Baumohl, "Inebriate Institutions in North America, 1840-1920," *British Journal of Addiction* 85 (1990): 1187-204; Edward M. Brown, "'What Shall We Do with the Inebriate?' Asylum Treatment and the Disease Concept of Alcoholism in the Late 19th Century," *Journal of the History of the Behavioral Sciences* 21 (1985): 48-59; Arnold Jaffe, *Addiction Reform in the Progressive Age* (New York: Arno Press, 1981).

50. "Conversation of Professor W. O. Atwater with Reverend Lyman Abbott, D.D., March 24 1898 . . ." (transcript), box 10, Atwater Papers; William H. Howell, *American Text-Book of Physiology* (Philadelphia: n.p., 1896), 298, quoted in Henry P. Bowditch and Clifton F. Hodge, "Report on the Present Instruction on the Physiological Action of Alcohol," in Billings, *Physiological Aspects,* 1: 11.

51. Harry G. Levine, "The Discovery of Addiction: Changing Conceptions of Habitual Drunkenness in America," *Journal of Studies on Alcohol* 39 (1978): 157; Jim Baumohl and Robin Room, "Inebriety, Doctors, and the State: Alcoholism Treatment Institutions before 1940," in *Recent Developments in Alcoholism* 5 (1987): 144-45. The notion of addiction had surfaced within the British medical community at least a century before Rush and other American physicians wrote about it. Their contribution was not to "discover" addiction, but to "psychiatrise" it—that is, to cast it as a disease that the asylum could rectify. Roy Porter, "The Drinking Man's

Disease: The Pre-History of Alcoholism in Georgian Britain," *British Journal of Addiction* 80 (1985): 385-96, esp. 393.

52. Baumohl, "Inebriate Institutions"; Leonard Blumberg, "The American Association for the Study and Cure of Inebriety," *Alcoholism: Clinical and Experimental Research* 2 (1978): 235-40.

53. "Teaching the Nature of Alcohol," *Quarterly Journal of Inebriety* 8 (1886): 177-78; L. D. Mason, "The Etiology of Dipsomania and Heredity of 'Alcoholic Inebriety,'" *Quarterly Journal of Inebriety* 10 (1888): 301-19. On heredity and alcoholism, see Bynum, "Alcoholism and Degeneration"; Charles E. Rosenberg, "The Bitter Fruit: Heredity, Disease, and Social Thought," in *No Other Gods: On Science and American Social Thought,* ed. Charles E. Rosenberg (Baltimore: Johns Hopkins University Press, 1976), 25-53.

54. "The New Year," *Quarterly Journal of Inebriety* 17 (1895): 68-69. For the shift from treatment to prevention within psychiatry, see Gerald Grob, *Mental Illness and American Society, 1875-1940* (Princeton: Princeton University Press, 1983), ch. 6.

55. "Teaching the Danger of the Use of Alcohol in Public Schools," *Quarterly Journal of Inebriety* 18 (1896): 291-92; Hunt, "Lecture Methods" (MS, n.d. [1883?]), frame 857, roll 13, STF Series.

56. *Quarterly Journal of Inebriety* 23 (1901): 94-95. On inebriests' increasing prominence in journals, meetings, and other professional activities, see "Lectures on Inebriety in Medical Colleges," *Quarterly Journal of Inebriety* 24 (1902): 342-44; Crothers to Hunt, 6 June 1902, 26 Aug. 1902, 8 Oct.1902, all in box 1, STF Papers.

57. WOA to MHH, 27 Oct. 1897, box 10, Atwater Papers; Sedgwick, "The Modern Subjection of Science," 52; L. D. Mason to MHH, 18 March 1903; A. H. Plumb to MHH, 22 Dec. 1903, both in box 1, STF Papers.

58. Mary L. Brumbach to MHH, 16 March 1904, frames 691-92, roll 2, STF Series; Bowditch and Hodge, "Report on the Present Instruction," 44; Mary Hunt, et al., *Reply of the Physiological Sub-Committee of the Committee of Fifty* (n.p., 1904).

59. Brumbach to MHH, 13 April 1904, frame 696, roll 2, STF Series.

60. Hunt, et al., *An Appeal to Truth;* "The Bureau of Scientific Temperance Investigation . . ." (MS, n.d. [1905?]), frames 288-92, roll 17, STF Series; Mary H. Hunt, "Scientific Temperance Instruction: Important Message to State, County, and Local Unions," *Union Signal,* 4 Feb. 1904. Thanks to the intercession of New Hampshire Senator Jacob Gallinger, a strong STI ally, 100,000 copies of Hunt's *Reply* were published as a U.S. Senate document. Thousands of copies were also mailed without postage under Gallinger's congressional frank. MHH, "Scientific Temperance Instruction," *Union Signal,* 14 April 1904.

61. Smith, *Co-Worker's Letter;* idem, *Objections to the Overton's Physiologies* (n.p., 1901), frame 241, roll 16, STF Series; MHH to Cora Stoddard, ? Dec. 1902, frames 124-25, roll 1, STF Series.

62. See, e.g., Mattie G. Shook, "Scientific Temperance Instruction," *Minutes of the Twenty-First Annual Meeting of the Woman's Christian Temperance Union of Tennessee* (Jonesboro: Herald and Tribune Print, 1902), 50-51; T. P. Bagby, "Scientific Temperance Instruction," *Minutes of the Twenty-Second Annual Convention of the Woman's Christian Temperance Union of Virginia* (Norfolk: S. B. Turner and Son, 1904), 49-50; MHH to Harriet B. Kells, 6 Oct. 1904, frame 266, roll 16, STF Series; *Report of the National Woman's Christian Temperance Union. Twenty-Eighth Annual Meeting* (Chicago: Woman's Temperance Publishing Association, 1901), 220.

63. See, e.g., E. O. Taylor to MHH, 22 Nov. 1901, frames 941-42, roll 2, STF Series; MHH to E. P. Hutchinson, 2 Dec. 1901, frame 678, roll 16, STF Series.

64. Eliza Mowry Bliven to MHH, 17 Nov. 1898, frames 690-92, roll 1, STF

Series. Under Connecticut's original 1886 STI measure, the state board of education published and distributed this "State Physiology." Although local school districts retained the right to select other texts, WCTU activists found the book so objectionable that they lobbied successively for a new law. As amended in 1893, the measure omitted all mention of a state textbook. *Minutes of the Woman's Christian Temperance Union of Connecticut. Seventeenth Annual Meeting* (Hartford: Press of the Connecticut Home, 1890), 38-39; *Report of the Commissioner of Education for the Year 1889-1890* (Washington, D.C.: Government Printing Office, 1891), 541.

65. "Report of the Committee on the Teaching of Hygiene in Public Schools," *Bulletin of the American Academy of Medicine* 7 (June 1905): 22; Joseph H. James to MHH, 17 June 1899, frame 717, roll 1, STF Series; "Survey by George Fitz," (MS, 1897), frame 64, roll 11, STF Series; MHH to A. C. Barnes, 11 Sept. 1901, frame 251, roll 16, STF Series; MHH to Mary A. Clark, 26 Sept. 1901, frames 264-65, roll 16, STF Series.

66. MHH to Mary F. Lovell, 21 Jan. 1902, frames 767-68, roll 15, STF Series; Sedgwick, "The Modern Subjection of Science," 52-53.

67. *Fifth Biennial Report of the Superintendent of Public Instruction, State of Minnesota* (St. Paul: Pioneer Press Co., 1888), 137-38; survey of L. J. Kehoo [Amsterdam, N.Y.], n.d. [1902], frame 931, roll 11, STF Series; survey of Mary F. Donaldson [Ellenville, N.Y.], n.d. [1902], frame 571, roll 12, STF Series. See also Mary H. Hunt, "The New Century and Scientific Temperance Instruction," *Union Signal*, 14 Feb. 1901; T. D. Crothers to Mary H. Hunt, 9 May 1901, box 1, STF Papers.

68. Mary H. Hunt, *Plan of Work of the Scientific Department of the National and International Woman's Christian Temperance Union* (Boston: n.p., 1888), 22; *Eighteenth Annual Report of the Woman's Christian Temperance Union of Michigan* (Grand Rapids: West Mill Printing Co., 1891), 51-53; "Mrs. Laitlaw's Method" (MS, 1890), enclosed with MHH to Mary Lovell, 7 Aug. 1890, frames 582-83, roll 16, STF Series.

69. *Fourteenth Annual Report of the Woman's Christian Temperance Union of Maryland* (Baltimore: Jas. M. Cumming, 1888), 23-24. See also Mary Allen West, "Scientific Temperance Instruction," *Minutes and Reports of the Woman's Christian Temperance Union of the State of Illinois, at the Seventeenth Annual Convention* (Chicago: Woman's Temperance Publishing Association, 1890), 135-40.

70. In 1886, for example, Hunt's critique of the much-despised "Smith's" textbook did not make a single reference to a specific set of data or experiments. Mary H. Hunt, "Why I Do Not Endorse Smith's Physiologies," *Union Signal*, 25 March 1886.

71. C. J. Smith to MHH, 20 April 1902, frames 923-24, roll 11, STF Series. See also Grace Bancroft Whitaker, "Scientific Temperance Instruction," *Report of the Twenty-Fourth Annual Convention of the Massachusetts Woman's Christian Temperance Union* (Boston: J. A. Cummings, 1897), 52-53.

72. MHH to Mary Lovell, 19 July 1890, frame 578, roll 13, STF Series; "The Connecticut Council of Education," *School Physiology Journal* 10 (March 1901): 99.

73. "Professor Atwater and Temperance Teaching," *Union Signal*, 15 March 1900; Hunt, et al., *Science the Arbiter*, 3; untitled editorial, *School Physiology Journal* 4 (March 1895): 84. See also Mary H. Hunt, "The New Century and Scientific Temperance Instruction," *Union Signal*, 14 Sept. 1901; untitled editorial, *Quarterly Journal of Inebriety* 22 (Jan. 1900): 103.

74. See Chapter Seven of this book for a full discussion of Hunt's complicated relationship with textbook publishers, who paid her royalties in exchange for her editorial work—and for her endorsement.

75. *Pennsylvania Medical Journal* 5 (1901-02): 613; Sedgwick, "The Modern

Subjection of Science," 52; Mary H. Hunt, "The Will of the People," *Report of the Commissioner of Education for the Year 1900-1901*, 2 vols. (Washington, D.C.: GPO, 1902), 1:1045.

76. *Journal of Proceedings and Addresses of the Annual Meeting of the National Educational Association* 39 (1900): 260; clipping, *Scranton Truth*, n.d. [1890?], frame 388, roll 14, STF Series; Mary H. Hunt, "The New Century Study of the Alcohol Question," *School Physiology Journal* 10 (Jan. 1901): 73.

77. MHH to Winfield S. Hall, 14 Sept. 1899, frames 78-79, roll 6, STF Series. On STI workers' household responsibilities, see, e.g., Josephine Weeks to MHH, 21 April 1886, frame 408, roll 3; E. M. Watson to MHH, n.d. [1886?], frames 418-19, roll 3; MHH to M. A. Barnard, 20 Oct. 1897, frame 623, roll 5, all in STF Series.

78. MHH, "The New Century Study," 73.

79. Mary H. Hunt, "Scientific Temperance Instruction," *Union Signal*, 21 Feb. 1901; idem, *An Epoch of the Nineteenth Century: An Outline of the Work for Scientific Temperance Education in the Public Schools of the United States* (Boston: P. H. Foster, 1897), 60; "The Bureau of Scientific Temperance Investigation . . . ," (MS, n.d. [1906?]), frame 288, roll 17, STF Series.

80. See, e.g., "Co-Worker's Letter," *School Physiology Journal* 7 (Jan. 1898): 6-8; MHH to Anna J. Lombard, 7 Nov. 1901, frames 179-81, roll 7, STF Series; West, "Scientific Temperance Instruction," 137-38.

81. Cora Stoddard to Jeannie Jones, 8 Nov. 1899, frame 657, roll 6, STF Series; Minutes of the Ann Arbor Woman's Christian Temperance Union, 14 Sept. 1899, box 1, Woman's Christian Temperance Union of Michigan Records, Bentley Library, University of Michigan, Ann Arbor, Michigan; Shook, "Scientific Temperance Instruction," 50-51.

82. L. P. Armstrong to Atwater, 11 Jan. 1900, box 5, Atwater Papers; *Thirtieth Annual Report of the Woman's Christian Temperance Union of Michigan* (Bay City: John P. Lambert, 1904), 41.

83. "Professor Atwater's Experiments," *Outlook* 62 (29 July 1899): 700-702; "The Temperance Text-Books," *Outlook* 66 (17 Nov. 1900): 706-9; "A Victory for Scientific Temperance," *Union Signal*, 7 March 1901; *Report of the Twenty-Eighth National WCTU Meeting*, 220.

4. "Let the People Decide"

1. Testimony of Charles R. Skinner before Senate and Assembly Committees on Education (MS, n.d. [1896]), frame ? (unintell.), roll 10, Temperance and Prohibition Papers, Ohio Historical Society (joint Ohio Historical Society—Michigan Historical Collections), Scientific Temperance Federation Series [hereafter STF Series]; Charles R. Skinner, "Place of Child-Study in a State System of Schools" (speech delivered to Illinois Society for Child-Study, Chicago, May 1896), folder 2, box 2, Charles Rufus Skinner Papers, Jefferson County Historical Society, Watertown, New York [hereafter Skinner Papers]. See also idem., "Temperate Education" (address to Mid-Year Conference of the New York State Woman's Christian Temperance Union, New York City, 6 May 1897), folder 2, box 1, Skinner Papers.

2. Albert H. Plumb, "Scientific Temperance Instruction in the Public Schools," *Report of the Commissioner of Education for the Year 1894-95*, 2 vols. (Washington, D.C.: Government Printing Office [hereafter GPO], 1896), 2: 1830-31.

3. L. D. Mason to MHH, 27 June 1903, box 1, Scientific Temperance Federation Papers, New York Public Library, New York, New York [hereafter STF Papers].

4. *The People Decide* (n.p., n.d. [1901?]), frame 210, roll 10, STF Series.

5. *Education* 5 (1885): 311-13; Frances Willard, "Lead Pencil Letter—No. III," *Union Signal,* 31 July 1884; Alice M. Guernsey, "New England Letter," *Union Signal,* 24 Dec. 1885; *Journal of Proceedings and Addresses of the National Education Association* [hereafter *NEA Proceedings*] 23 (1884): 15; Mary H. Hunt, *A History of the First Decade of the Department of Scientific Temperance Instruction . . . ,* 2nd ed. (Boston: George E. Crosby, 1891), 40-42; John Q. Stewart, "Temperance in Its Relation to Education," *Pennsylvania School Journal* 35 (1886): 90-91. On Bicknell, see David Tyack and Elisabeth Hansot, *Managers of Virtue: Public School Leadership in America, 1820-1980* (New York: Basic, 1982), 99; for other evidence of schoolman support during these years, see Mary Hunt, "Report of the Superintendent of the Department of Scientific Temperance Instruction" (MS, 1881), box 1, STF Papers; "A Good Law," *Iowa Normal Monthly* 9 (March 1886): 312; n.t., *Michigan School Moderator* 6 (18 October 1885): 53.

6. "Discussion," *Pennsylvania School Journal* 34 (Sept. 1885): 123; *Reports of the School Commissioner of the City of Springfield, for 1887* (Springfield, Mass.: Weaver, Shipman, and Co., 1888), 53; clipping, n.p, n.d. [1884?], frame 54, roll 8, STF Series. See *Fiftieth Annual Report of the Superintendent of Public Instruction of the State of Michigan* (Lansing: Thorp and Godfrey, 1887), iii-v; *Report of the Superintendent of Public Instruction of the Commonwealth of Pennsylvania* (Harrisburg: Edwin K. Meyers, 1888), x.

7. "Alcohol in Schools," *Journal of Education,* 8 Jan. 1885, frame 407, roll 17, STF Series; Mary Allen West, "Temperance Education," *Illinois School Journal* 4 (April 1885): 632; *Education* 5 (1885): 311; Carl Kaestle, *Pillars of the Republic: Common Schools and American Society, 1780-1860* (New York: Hill and Wang, 1984), 115.

8. Robert Wiebe, "The Social Functions of Public Education," *American Quarterly* 21 (1969): 147-64; Tyack and Hansot, *Managers of Virtue,* pt. 2; Chicago Merchants' Club, *Public Schools and Their Administration: Addresses Delivered at the Fifty-Ninth Meeting of the Merchants' Club of Chicago* (Chicago: Merchants' Club, 1906), 40, quoted in David Tyack, *The One Best System: A History of American Urban Education* (Cambridge: Harvard University Press, 1974), 77.

9. Tyack and Hansot, *Managers of Virtue,* 3; John Higham, "Hanging Together: Divergent Unities in American History," *Journal of American History* 61 (1974): 24, 15; Robert Wiebe, *The Search for Order, 1877-1920* (New York: Hill and Wang, 1967), 159-63; Richard L. McCormick, *From Realignment to Reform: Political Change in New York State, 1893-1910* (Ithaca: Cornell University Press, 1981), 255.

10. Andrew S. Draper, "Common Schools in the Larger Cities," *The Forum* 27 (1899): 385-97, quoted in Tyack, *One Best System,* 130; McCormick, *From Realignment to Reform,* 255; Tyack and Hansot, *Managers of Virtue,* 129-67.

11. Tyack and Hansot, *Managers of Virtue,* 204.

12. U.S. Bureau of Education, *Report of the Committee on Secondary School Studies . . .* (Washington, D.C.: GPO, 1893), 138; William T. Harris, "Report of the Sub-Committee on the Correlation of Studies in Elementary Education," *NEA Proceedings* 34 (1895): 310-11; *Report of the Commissioner of Education for the Year 1895-1896,* 2 vols. (Washington, D.C.: GPO, 1897), 2:1361; *NEA Proceedings* 39 (1900): 250-66.

13. *Fifteenth Annual Report of the Woman's Christian Temperance Union of Illinois* (Chicago: Woman's Temperance Publishing Association, 1888), 92; *Report of the Eighteenth Annual Meeting of the Woman's Christian Temperance Union of the State of New York* (Oswego: R. J. Oliphant, 1889), 76.

14. In North Dakota, for example, a woman state superintendent criticized STI's assumption that "scientific facts" would influence children's "conduct in life." Meanwhile, Colorado's female superintendent charged that schools were overloading young minds with STI. Rejecting the WCTU's call for annual instruction in the subject, she

suggested a streamlined, two-year course of study. In Mary Hunt's native Massachusetts, finally, education secretary and former Wellesley College president Alice Freeman Palmer spearheaded an unsuccessful drive to weaken the state's STI law. *Fourth Biennial Report of the Superintendent of Public Instruction, to the Governor of North Dakota* (Jamestown: Alert, 1894), 17-18; *Twelfth Biennial Report of the Superintendent of Public Instruction of the State of Colorado* (Denver: Smith-Brooks, 1900), 608-9; "Scientific Temperance," *Journal of Education* 23 (March 1899), frame 419, roll 14, STF Series; MHH to Katharine Stevenson, 18 Jan. 1900, frames 346-47, roll 17, STF Series. On female superintendents in the West, see Tyack and Hansot, *Managers of Virtue*, 187.

15. "Report of the Committee on Scientific Temperance Instruction," (MS, n.d. [1901?]), Burlington WCTU, Woman's Christian Temperance Union of Vermont Records, Bailey-Howe Library, University of Vermont, Burlington, Vermont. For other examples of STI opposition among local schoolwomen, see, e.g., "Scientific Temperance Instruction in Virginia," *Union Signal*, 2 May 1901; "Temperance Teaching in the Public Schools: A Symposium," *Connecticut Citizen* 8 (Oct. 1901): 1-2.

16. W. B. Ferguson, "Temperance Teaching and Recent Legislation in Connecticut," *Educational Review* 23 (1902): 249; *Forty-Second Annual Report of the Superintendent of Public Instruction of the State of New York* [1894-95] (Albany: Wynkoop Hallenback Crawford Co., 1896), xlii; *Protest against the Bill Unnecessarily Increasing Physiology and Hygiene with Special Reference to the Nature of Alcoholic Drinks and Other Narcotics* (n.p., 1895), 112; *Educational Review* 11 (1896): 307; Massachusetts Board of Education, *Sixty-Third Annual Report of the Board of Education . . .* [1898-99] (Boston: Wright and Potter, 1900): 568. See also *NEA Proceedings* 39 (1900): 254; *School Bulletin* [N.Y.] 22 (Dec. 1895): 51; *Auburn* [N.Y.] *Bulletin*, 4 June 1895; "Trouble over the New Law," *New York Daily Tribune*, 24 Jan. 1896, p. 14.

17. *Report of the Commissioner of Education for the Year 1895-96* 2:1361; *Report of the Superintendent of Public Instruction of the Commonwealth of Pennsylvania* (Harrisburg: Wm. Stanley Ray, 1900), 50; W. B. Ferguson, "Scientific Temperance in Connecticut," *Journal of Education*, 13 June 1901, frame 712, roll 14, STF Series; Wilbur O. Atwater, "Rational Temperance Reform" (MS, n.d. [1900?]), box 10, Wilbur O. Atwater Papers, Olin Library Wesleyan University, Middletown, Connecticut [hereafter Atwater Papers]; *NEA Proceedings* 39 (1900): 262-63. See also "Temperance Teaching in the Schools," *New York Times*, 9 Jan. 1896. On women reformers in the schools, see William J. Reese, *Power and the Promise of School Reform: Grassroots Movements during the Progressive Era* (Boston: Routledge and Kegan Paul, 1986), ch. 2; Michael W. Sedlak, "Attitudes, Choices, and Behavior: School Delivery of Health and Social Services" in *Learning from the Past: What History Teaches Us about School Reform*, ed. Diane Ravitch and Maris A. Vinovskis (Baltimore: Johns Hopkins, 1995), 58-67.

18. Dorothy Ross, *G. Stanley Hall: The Psychologist as Prophet* (Chicago: University of Chicago Press, 1972), ch. 15; Herbert M. Kliebard, *The Struggle for the American Curriculum, 1893-1958* (Boston: Routledge and Kegan Paul, 1986), 41-51; Richard Hofstadter, *Anti-Intellectualism in American Life* (New York: Vintage, 1962), ch. 14; Joseph F. Kett, *Rites of Passage: Adolescence in America, 1790 to the Present* (New York: Basic, 1977), ch. 7.

19. *Union Signal*, 19 Aug. 1897; Ferguson, "Temperance Teaching," 242, 235; *Report of the Superintendent of Public Instruction of the Commonwealth of Pennsylvania* (Harrisburg: Wm. Stanley Ray, 1898), x. See also Massachusetts Board of Education, *Fifty-Eighth Annual Report of the Board of Education . . .* [1893-94] (Boston: Wright and Potter, 1895), 138-40.

20. Leila Zenderland, "Education, Evangelism, and the Origins of Clinical Psychol-

ogy: The Child-Study Legacy," *Journal of the History of the Behavioral Sciences* 24 (April 1988): 152; Kliebard, *Struggle for the American Curriculum*, 44-46; Ross, *G. Stanley Hall*, 314-15; *NEA Proceedings* 39 (1900): 253; Skinner, "Temperate Education."

21. As David Tyack and Elisabeth Hansot have noted, small-town superintendents during this era often "acquired new ideas about education in their training and in their professional associations" and became "carriers of an adopted cosmopolitanism." Tyack and Hansot, *Managers of Virtue*, 178.

22. Frank E. Parlin to R. C. Magwood, 4 April 1904, frame 761, roll 12, STF Series; F. C. Atwell to Magwood, 5 April 1904, frames 740-41, roll 12, STF Series; "Temperance Teaching in the Public Schools. A Symposium," 1-2. See also Charles E. Stevens to R. C. Magwood, 23 April 1904, roll 12, frames 764-65, STF Series; *Report of the Superintendent of Public Instruction of the Commonwealth of Pennsylvania* (Harrisburg: Wm. Stanley Ray, 1900), 50.

23. Kaestle, *Pillars of the Republic*, 87-88; Mary Ryan, *Cradle of the Middle Class: The Family in Oneida County, New York, 1790-1865* (Cambridge, Eng.: Cambridge University Press, 1981), 99-101; Paul E. Johnson, *A Shopkeeper's Millennium: Society and Revivals in Rochester, New York, 1815-1837* (New York: Hill and Wang, 1978), ch. 5; James H. Moorhead, "Between Progress and Apocalypse: A Reassessment of Millennialism in American Religious Thought, 1800-1880," *Journal of American History* 71 (1984): 525.

24. Susan Curtis, *A Consuming Faith: The Social Gospel and Modern American Culture* (Baltimore: Johns Hopkins University Press, 1991), 5, 8; Ferenc Morton Szasz, *The Divided Mind of Protestant America, 1880-1930* (University: University of Alabama Press, 1982), 16, 40. Over half of the leading liberal ministers in the 1890s received some training in Germany. William R. Hutchinson, *The Modernist Impulse in American Protestantism* (Cambridge: Harvard University Press, 1976), 122.

25. Curtis, *Consuming Faith*, 150; Hutchinson, *Modernist Impulse*, 116; "Professor Atwater's Experiments," *Outlook* 62 (29 July 1899): 700; "Temperance Textbooks," *Outlook* 66 (22 Dec. 1900): 975; "The Temperance Text-Books," *Outlook* 66 (17 Nov. 1900): 707; "An Appeal to Truth," *Outlook* 64 (17 Feb. 1900): 390-91.

26. "Bishops Opposed to It," *New York Daily Tribune*, 2 June 1895, p. 10; "Professor Atwater's Experiments," 700; "The Temperance Text-Books," 707-8. Abbott submitted at least one of his anti-STI editorials to Atwater before its publication, asking him to return it "with any suggestions." Abbott to Atwater, 31 January 1900, box 5, Atwater Papers.

27. *Protest against the Bill*, 27; Nicholas M. Butler to Atwater, 8 Nov. 1899, box 4; S. T. Dutton to Atwater, 13 Nov. 1899, box 5; Joseph Anderson to Atwater, 13 April 1900, box 5; Dutton to Atwater, 20 March 1900, box 5, all in Atwater Papers.

28. For a complete discussion of these challenges and Hunt's campaigns to rebut them, see Chapter Six of this book.

29. "Scientific Temperance," *Journal of Education*, 23 March 1899, frame 419, roll 14, STF Series; *Report of the Twenty-Seventh Annual Meeting of the Woman's Christian Temperance Union of the State of New York* (Ithaca, N.Y.: Press of the Ithaca Journal, 1900), 168.

30. M. P. Cooley to Cora Stoddard, 17 May 1901, frames 903-5, roll 14, STF Series.

31. Kliebard, *Struggle for the American Curriculum*, 18-20; Merle Curti, *The Social Ideas of American Educators* (1935; Paterson, N.J.: Pageant Books, 1959), ch. 9; Selwyn K. Troen, *The Public and the Schools: Shaping the St. Louis System, 1838-1920* (Columbia: University of Missouri Press, 1975), ch. 8, esp. 159.

32. W. T. Harris, "The Commissioner's Introduction," in *Report of the Commissioner*

of Education for the Year 1900-1901, 2 vols. (Washington, D.C.: GPO, 1902), 1: xlii; W. T. Harris, "How to Use Text-Books Effectively," *School Physiology Journal* 16 (Oct. 1906): 17-19.

33. On the "short-lived heyday" of child-study in the 1890s, see Kliebard, *Struggle for the American Curriculum,* 18; Kett, *Rites of Passage,* 228; Ross, *G. Stanley Hall,* 358-59.

34. Curti, *Social Ideas of American Educators,* 343; *School Bulletin* [N.Y.] 22 (Oct. 1895): 28; *New York Times,* 17 June 1895; *Protest against the Bill,* 104-5; "House Day," *Cincinnati Enquirer,* 2 March 1888.

35. Kliebard, *Struggle for the American Curriculum,* xi. In a more recent edition of his book, Kliebard identifies these four camps as "reform sub-groups." Kliebard, *Struggle for the American Curriculum, 1893-58,* 2nd ed. (Boston: Routledge, 1995), 243. I prefer his new language to "interest groups," which implies—erroneously— that the reformers enlisted large lay constituencies in their respective causes.

36. William T. Sedgwick, "The Modern Subjection of Science and Education to Propaganda," *Science* 15 (10 Jan. 1902): 51. See also George Groff to William T. Harris, 17 March 1894, frame 453, roll 9, STF Series.

37. William A. Mowry, *Recollections of a New England Educator, 1838-1908* (New York: Silver, Burdett, 1908), 154-55; National Education Association, *Report of the Committee of Fifteen on Elementary Education . . .* (New York: American Book Co., 1895), 144; *Proceedings of the School Committee of the City of Boston, 1891* (Boston: Rockwell and Churchill, 1891), 20-21; William A. Mowry, "Powers and Duties of School Superintendents," *Educational Review* 9 (1895): 47; *Biennial Report of the Superintendent of Public Instruction of the State of Iowa* [1894-95] (Des Moines: F. A. Conaway, 1895), 85-86; *Protest against the Bill,* 27-28.

38. Cora Stoddard, "Moral or Scientific Temperance Instruction," *School Physiology Journal* 20 (Oct. 1910): 26; Hunt, *History of the First Decade,* 41; Mary H. Hunt, "Temperance Instruction in the Lower Grades," *School Physiology Journal* 5 (Oct. 1895): 18.

39. "To What Extent and by What Methods Should Temperance Physiology Be Taught in the Public Schools?" *School Physiology Journal* 11 (Nov. 1901), extra ed., p. 4. For similar invocations of child-study rhetoric, see "WCTU Work in Public Schools," (MS, n.d. [1897?]), pp. 6-7, folder 20, Catharine Dunham Papers, Arthur and Elizabeth Schlesinger Library, Radcliffe College, Cambridge, Massachusetts.

40. Mary Hunt, "How to Make Room," *School Physiology Journal* 4 (March 1895): 80. In his own battle with child-study critics, William T. Harris would also appropriate—and, his critics charged, misinterpret—the "correlation" concept. Kliebard, *Struggle for the American Curriculum,* 19.

41. *Thirty-Second Annual Report of the State Superintendent of Common Schools. State of Maine* [1885] (Augusta: Sprague and Son, 1886), 46; Ross, *G. Stanley Hall,* 312-13. For other STI barbs at the rigidity and formalism of the traditional curriculum, see *Tenth Annual Report of the Woman's Christian Temperance Union of Illinois* (Chicago: Union Signal Print, 1883), 21; "Do You 'Care'?" *School Physiology Journal* 6 (Sept. 1896): 9.

42. "Compulsory Temperance Education," *Hartford Courant,* 6 April 1901; S. G. Richards, "Physiology in the Public Schools," *Iowa Normal Monthly* 23 (1900): 403-4; George H. Talbot, "Physiology," *Colorado School Journal* 5 (July 1889): 3. See also Oscar Gerson, "Can We Dispense with Physiology?" *The Teacher* [Philadelphia] 2 (1903): 210-11; W. S. Hall, "The Value of Instruction Regarding Alcohol," *School Physiology Journal* 16 (Dec. 1906): 49-51.

43. Mary Hunt, "Is It a Moral or a Scientific Question, or Both?" *North Carolina*

White Ribbon (April 1900), frame 357, roll 17, STF Series; Mary Hunt, "These Are the Commandments," *School Physiology Journal* 6 (Jan. 1897): 84–85.

44. Macy A. Smith, "Some Untenable Criticisms on the Temperance Education Laws," *School Physiology Journal* 12 (Feb. 1903): 92–93; "Corresponding Secretary's Notes," *Union Signal*, 4 Jan. 1900; *Report of the Thirty-First Annual Meeting of the Woman's Christian Temperance Union of the State of New York* (Newburgh: Journal Printing House and Book-Binding, 1904), 44–45. See also J. N. David, "Temperance, Text Books and Teachers," *West Virginia School Journal* 6 (Feb. 1887): 11; A. E. Winship, "Temperance Physiology," *School Physiology Journal* 9 (April 1900), n.p.; Susanna M. D. Fry, "Defense of Scientific Temperance Instruction," *Union Signal*, 20 March 1902.

45. *Report of the Thirty-First Meeting of the WCTU of New York*, 44–45; "Discussion," *NEA Proceedings* 39 (1900): 256; Plumb, "Scientific Temperance Instruction in the Public Schools," 1830; Mary H. Hunt, *An Epoch of the Nineteenth Century: An Outline of the Work for Scientific Temperance Education in the Public Schools of the United States* (Boston: P. H. Foster, 1897), 29; "Those Temperance Text-Books," *Outlook* 63 (28 Nov. 1899): 485; Walter Lippmann, *A Preface to Politics* (New York: Macmillan, 1933), 120–21.

46. Mary Hunt, "A Great Victory for a Stronger Scientific Temperance Law in Illinois," *Union Signal*, 24 June 1897; clipping, *School Physiology Journal*, Sept. 1895, frame 217, roll 10, STF Series. For background on Hunt's minister allies, see the following articles in Henry Warner Bowden, *Dictionary of American Religious Biography* (Westport, Conn.: Greenwood, 1977): "Cook, Joseph," 111–12; "Thompson, Charles Lemuel," 549–50; "Clark, Francis Edward," 113–14; "Dixon, Amzi Clarence," 148–49.

47. Hutchinson, *Modernist Impulse*, 116; Francis Peabody to Hunt, 11 Oct. 1904, frames 204–5, roll 3, STF Series; George Fitz to Henry P. Bowditch, ? Feb. 1899, frame 138, roll 11, STF Series; Richard W. Boynton to George C. Lorimer, 2 March 1899, frames 631–32, roll 15, STF Series; "Discussion," *NEA Proceedings* 39 (1900): 255; John Fiske, "Guessing at Half and Multiplying by Two" in *A Century of Science* (Boston: Houghton, Mifflin, 1900), 336, 344–45. See also "Lawrence, William," *National Cyclopaedia of American Biography* (New York: James T. White and Co., 1930), Current Volume C: 479; "Hodges, George," *National Cyclopaedia of America Biography* (New York: James T. White and Co., 1932), 22: 79–80. Hunt apparently tried to enlist both Lawrence and Hodges on behalf of STI, but both ministers turned her down. Fitz to Bowditch, ? Feb. 1899, frame 138, roll 11, STF Series.

48. Fiske, "Guessing at Half," 333; Massachusetts Board of Education, *Sixty-Third Annual Report*, 543–49; *Documents Printed by the Order of the Senate of the Commonwealth of Massachusetts*, no. 282 (n.p., 1899).

49. To bring its schools into compliance with the 1884 law, for example, New York City required just one twenty-minute STI lecture each month. Rochester declared that a single lesson per term would suffice, while Syracuse seems to have ignored STI altogether. *Forty-Fourth Annual Report of the Board of Education of the City of New York* [1885] (New York: Hall of the Board of Education, 1886), 25; Rochester Public Schools, *Thirty-Seventh Annual Report of the Board of Education* [1883–84] (Rochester: Sunday Herald Job Office, 1884), 110; *Thirty-Seventh Annual Report of the Board of Education of the City of Syracuse* [1884–85] (Syracuse: Standard Book and Job Print, 1884).

50. Hunt, *Epoch of the Nineteenth Century*, 9; *Protest against the Bill*, 113–14 and passim; *Laws of the Forty-Four States of the Union on the Teaching of Scientific Temperance* (Albany: Weed-Parsons, 1895), 1; "Cattaraugus," *School Bulletin* 22 (Feb. 1896): 97.

On the Public Education Association, see Diane Ravitch, *The Great School Wars: New York City, 1805-1973* (New York: Basic, 1974), ch. 14, and Sol Cohen, *Progressives and Urban School Reform: The Public Education Association of New York City, 1895-1954* (New York: Teachers College, 1964), chs. 1-2.

51. "Cattaraugus," 97; Ravitch, *Great School Wars,* 155.

52. *Open Letter,* (n.p., 24 Jan. 1895), frame 197, roll 10, STF Series; Hunt to Samuel L. Beiler, 21 Nov. 1895, folder AC 5810-1, American University Presidential Papers, American University Library, Washington, D.C.; Mary E. Newton, "A Woman's Plea for the Ainsworth Law," *Brooklyn Daily Eagle,* 24 March 1896. The New York Central Committee for Scientific Temperance Instruction in Public Schools included over twenty-six ministers from at least a dozen different denominations. For a list, see *Open Letter,* frame 196.

53. Hunt, *Epoch of the Nineteenth Century,* 24-25; "Signed on the Last Day," *New York Daily Tribune,* 16 June 1895; Skinner to Frederick W. Holls, 22 June 1895, box 7, Frederick W. Holls Papers, Butler Library, Columbia University; *Public Papers of Levi P. Morton . . .* (Albany: Weed-Parsons, 1896), 138, 132. On efforts by both sides to sway Morton, see "Temperance in Schools," *New York Times,* 7 June 1895; "Vigorous Protest Made," ibid., 8 June 1895; "Temperance in Schools," ibid., 10 June 1895. Owner of a restaurant that served liquor, Morton was long derided by Republican prohibitionists as "Rumseller Morton." Signing the STI measure promised to appease these critics without alienating the wet, antiprohibitionist wing of his party. See Richard F. Hamm, *Shaping the Eighteenth Amendment: Temperance Reform, Legal Culture, and the Polity, 1880-1930* (Chapel Hill: University of North Carolina Press, 1995), 29; McCormick, *From Realignment to Reform,* 61-63.

54. See, e.g., *Central Committee's Reply* (n.p., June 1895), frame 210, roll 10, STF Series; *School Physiology Journal* 9 (May 1900): 139.

55. On churches in the Illinois campaign, see Hunt, *Epoch of the Nineteenth Century,* 30-33; Mary H. Hunt, "A Great Victory for a Stronger Scientific Temperance Law in Illinois," *Union Signal,* 24 June 1897. In Massachusetts, see *Report of the Temperance Committee for the Massachusetts State Association of Congregated Churches* (n.p., 18 May 1898), frames 405-8, roll 14, STF Series; *The Truth About the Proposed Amendment to the Temperance Education Law* (n.p., n.d. [1898]), frames 413-15, roll 14, STF Series.

56. A. Plumb, "Professors Sedgwick and Hough on Temperance Instruction in the Public Schools," *Boston Transcript,* 20 May 1903, frame 389, roll 17, STF Series; "Temperance Instruction in the Schools," *School Physiology Journal* 8 (March 1899), extra ed.; *The People Decide.*

57. "A Question of Majorities," *School Physiology Journal* 10 (Oct. 1900): 25; Mary H. Hunt, "The Present Need and How to Meet It," *Union Signal,* 24 April 1901; MHH to Cornelia Forbes, 6 Dec. 1899, frame 9, roll 17, STF Series. See also MHH to W. T. Harris, 24 Oct. 1901, frames 411-12, roll 7, STF Series.

58. E. O. Orr to MHH [1902?], frame 383, roll 12, STF Series; *Minutes of the 10th Annual Convention of the Arizona Woman's Christian Temperance Union* (Phoenix: Gazzette Job Office Printers, 1898), 14; Mary E. Newton, "A Woman's Plea for the Ainsworth Law," *Brooklyn Daily Eagle,* 24 March 1896, p. 11.

59. Mary E. Tallmann to MHH, 6 Feb. 1902, frame 65, roll 12, STF Series; Newton, "A Woman's Plea," 11.

60. "Bad Faith is Charged," *New York Times,* 22 March 1896. A prominent figure in state and national reform circles, Maxwell had cofounded the *Educational Review* with Nicholas Murray Butler. When Brooklyn united with Manhattan and three other

boroughs to form Greater New York in 1898, Maxwell would become its first school superintendent. Ravitch, *Great School Wars*, 163-64.

61. "A Breezy WCTU Meeting," *Brooklyn Daily Eagle*, 25 March 1896; Newton, "A Woman's Plea," 11.

62. Trained as a teacher, Maxwell was an instructor in the Brooklyn evening schools before moving into administration. Ravitch, *Great School Wars*, 164. On opposition from the State Teachers' Association, see, e.g., *New York Times*, 3 July 1895; *Protest Against the Bill*, 116-17; "Convention Over," *Rochester Daily Union*, 27 September 1895.

5. THE DILEMMA OF MISS JOLLY

1. "In Perplexity," *Ohio Educational Monthly* 40 (March 1891): 122-23.

2. Margaret L. Macready, "Temperance Instruction for Little Ones," *Ohio Educational Monthly* 10 (July 1891): 315-17; letter from Hattie W. Wetmore, *Ohio Educational Monthly* 10 (July 1891): 317-18.

3. Mary Lovell to George Groff, 5 Feb. 1892, frames 250-51, roll 9, Temperance and Prohibition Papers, Ohio Historical Society (joint Ohio Historical Society—Michigan Historical Collections), Scientific Temperance Federation Series [hereafter STF Series]; Mary H. Hunt, "How She Met Current Objections" (MS, 1899), frame 880, roll 13, STF Series.

4. This is the theme of Marjorie Murphy, *Blackboard Unions: The AFT and the NEA, 1900-1980* (Ithaca, N.Y.: Cornell University Press, 1990). For a similar account, see David Hogan, *Class and Reform: School and Society in Chicago, 1880-1930* (Philadelphia: University of Pennsylvania Press, 1985), ch. 5.

5. Murphy, *Blackboard Unions*, 2n.

6. "Temperance in Schools," *Michigan School Moderator* 5 (18 Sept. 1884): 46.

7. MHH to Samuel L. Beiler, 21 July 1894, folder AC 5810-1 ["Proposal for College of Scientific Temperance 1894-98"], American University Presidential Papers, American University Library, Washington, D.C. [hereafter AU Papers].

8. Joseph Mayer Rice, *The Public School System of the United States* (New York: Arno Press, 1969 [1893]), 60-61.

9. "Teach Pupils How to Study," *School Physiology Journal* 6 (March 1897): 121; "Where Will the Responsibility Lie?" *School Physiology Journal* 6 (Jan. 1897): 73; untitled editorial, *School Physiology Journal* 6 (Oct. 1896): 29-30.

10. "Temperance Physiology for Little Folks," *Michigan School Moderator* [hereafter *MSM*] 6 (8 Oct. 1885): 63-64; "Temperance Physiology," *MSM* 6 (22 Oct. 1885): 86; "Temperance Physiology for the Little Folks," *MSM* 6 (3 Dec. 1885); "Temperance Physiology," *MSM* 6 (7 Jan. 1886): 209; "Temperance Physiology," *MSM* 6 (15 April 1886): 367; "Temperance Physiology," *MSM* 6 (6 May 1886): 390.

11. Rice, *Public School System*, esp. ch. 9; Barbara Finkelstein, *Governing the Young: Teacher Behavior in Popular Primary Schools in the Nineteenth-Century United States* (New York: Falmer, 1989), 87; Larry Cuban, *How Teachers Taught: Constancy and Change in American Classrooms, 1890-1980* (New York: Longman, 1984), 24.

12. "Reporter," "Oswosso Schools," *MSM* 5 (26 Feb. 1885): 468; *MSM* 5 (19 March 1885): 529.

13. G. V. Whaley, "Physiology as Taught and as It Should Be Taught," *Iowa Normal Monthly* 19 (March 1896): 380-81; C. F. Warner, "Scientific Temperance Teaching," *Thirty-Fourth Annual Report of the State Superintendent of Common Schools.*

State of Maine [1887] (Augusta: Burleigh and Flynt, 1888), 123-32; "Experimental Lessons on Alcohol," ibid., 84-96; "Lesson on Alcohol and the Stomach," *MSM* 4 (24 April 1884): 604; *Biennial Report of the State Superintendent of Public Education* . . . [Louisiana, 1900-1901] (Baton Rouge: The Advocate, 1902), 93-94; Cassie R. Spencer, "Lesson in Hygiene," *New Mexico Journal of Education* 15 (Feb. 1919): 7-9.

14. Whaley, "Physiology as Taught"; remarks of C. E. Gorton [Yonkers], 1884 Minutes, box 1, Records of the Council of School Superintendents, Cities and Villages, State of New York, Department of Manuscripts and University Archives, Cornell University Library, Ithaca, New York [hereafter Council of School Superintendents].

15. Warner, "Scientific Temperance Teaching"; John Madden to MHH, 26 Dec. 1902, box 1, Scientific Temperance Federation Papers, New York Public Library, New York, New York [hereafter STF Papers].

16. Cora Lee Johnson, "Prominence of Temperance Instruction. What, How and How Much Should Be Taught in the Grades Concerning Narcotics and Alcoholism?" *New Mexico Journal of Education* 5 (15 Feb. 1909): 53.

17. Jane I. Newell, "The Woman's Christian Temperance Union in America" (Ph.D. diss., University of Wisconsin, 1919), 94; "The Woman's Christian Temperance Union and Education," *School Physiology Journal* 12 (Nov. 1902): 36; John L. Rury, "Who Became Teachers? The Social Characteristics of Teachers in American History," in *American Teachers: Histories of a Profession at Work*, ed. Donald Warren (New York: Macmillan, 1989), 10.

18. See. e.g., *Report of the Thirtieth Annual Meeting of the Woman's Christian Temperance Union of the State of New York* (Newburgh: Journal Printing House and Book-Binding, 1903), 136; *Eighth Annual Report of the Woman's Christian Temperance Union of Pennsylvania* [1882] (Philadelphia: Lineaweaver and Wallace, 1883), 57, 63; *Twelfth Annual Meeting of the Woman's Christian Temperance Union of Iowa* (Dubuque: Telegraph-Herald Printers, 1901), 77.

19. David K. Cohen, "Teaching Practice: Plus Que Ca Change . . . ," in *Contributing to Educational Change: Perspectives on Research and Practice*, ed. Phillip W. Jackson (Berkeley: McCutchan, 1988), 44-47; Geraldine Joncich Clifford, "Man/Woman/ Teacher: Gender, Family, and Career in American Educational History" in Warren, ed., *American Teachers*, 315; Wayne E. Fuller, *The Old Country School: The Story of Rural Education in the Middle West* (Chicago: University of Chicago Press, 1982), 201-7; Wayne E. Fuller, "The Teacher in the Country School," in Warren, ed., *American Teachers*, 107-8; Jennie Chapman, "The School Ma'am's Dilemma," *MSM* 5 (26 Feb. 1885): 465; "Quo Vadimus," *New York Education* 3 (Oct. 1899): 82.

20. Jeffrey Glanz, *Bureaucracy and Professionalism: The Evolution of Public School Supervision* (Rutherford, N.J.: Fairleigh Dickinson University Press, 1991), 37-76; Murphy, *Blackboard Unions*, ch. 2; Fuller, *Old Country School*, 192; Floyd A. Raze, "Reflections of a School Teacher on Being Visited by the Commissioner," *MSM* 21 (20 Dec. 1900): 236-37.

21. *Minutes of the Thirteenth Annual Convention of the Woman's Christian Temperance Union of the State of Minnesota* (Minneapolis: Greenwood and Souther, 1889), 77-78; "Why Use the Visitation Blank and How," *School Physiology Journal* 7 (Jan. 1898): 6; Glanz, *Bureaucracy and Professionalism*, 61-63; MHH to J. H. Gardner, 9 Nov. 1897, frame 804, roll 5, STF Series.

22. MHH to Mrs. George E. Deuel, 25 Sept. 1901, frame 53, roll 7, STF Series; Woman's Christian Temperance Union, *The Pathfinder: or National Plans for Securing Scientific Temperance Education in Schools and Colleges* (New York: A. S. Barnes, 1885), 15; MHH to M. Barley, 9 Oct. 1901, frame 242, roll 7, STF Series; *Union Signal*, 16 Aug. 1894; *Minutes of the Twenty-Third Annual Convention of the Woman's Chris-*

tian Temperance Union of Virginia (n.p., 1905), 63; ? to A. K. Whitcomb, 16 Jan. 1902, frame 4, roll 3; Ethel Turner to Cora Stoddard, 13 Feb. 1923, frames 339-40, roll 22, both in STF Series.

23. "Report of the Committee on the Teaching of Hygiene in Public Schools," *Bulletin of the American Academy of Medicine* 7 (June 1905): 8; N. E. F., "Jottings from Byron Center," *MSM* 5 (7 May 1885): 649; *Annual Report of the Superintendent of Public Instruction, Being the Thirty-Eighth Annual Report upon the Public Schools of New Hampshire* (Concord: Parson B. Cogswell, 1884), 57; *Report of the Commissioner of Education for the Year 1893-94*, 2 vols. (Washington, D.C.: Government Printing Office, 1895), 2:1159.

24. *Minutes of the Fourteenth Annual Meeting of the Woman's Christian Temperance Union of the State of Minnesota* (Minneapolis: Housekeeper Print, 1890), 100-101; MHH to J. G. Trurgn, 31 Oct. 1899, frames 567-68, roll 6, STF Series.

25. Robert Bader, *Prohibition in Kansas* (Lawrence: University of Kansas Press, 1986), 100; *Fourteenth Annual Meeting of the Woman's Christian Temperance Union of Iowa* (Cedar Rapids: *Daily Republican*, 1887), 33; MHH to Rebecca A. Brown, 11 November 1897, frame 870, roll 5, STF Series.

26. M. McDuffee to Mary Newton, 29 Sept. 1897, frame 312; McDuffee to Mary Lovell, 6 Oct. 1897, frame 400; McDuffee to Hannah Whitson Lovell, 16 Oct. 1897, frame 551, all in roll 5, STF Series; MHH to L. Ellen Day, 15 Sept. 1899, roll 6, STF Series.

27. "How," a Hunt ally asked, "can teachers be expected to feel an interest in teaching a subject which they have not mastered?" Axel Gustafson, "Temperance Teaching in Schools," in *Fifty-Seventh Annual Meeting of the American Institute of Instruction* (Boston: Willard Small, 1887), 171-73.

28. D. Bentley, "The Responsibility of Mothers in Regard to Temperance Teaching in Our Schools," *Union Signal*, 28 April 1887; Alice J. White, "To Teachers and Others," *Our Message Supplement* (1889), frame 304, roll 17, STF Series.

29. *Biennial Report of the Superintendent of Public Instruction of the State of Iowa* [1886-87] (Des Moines: George E. Roberts, 1887), 75; survey of Alida Johnson [Brooklyn], frame 785, roll 11, STF Series.

30. Survey of Mrs. J. B. Story [Perth, N.Y.], frame 707, roll 11; survey of ?, frame 238, roll 11; survey of Mary A. Sayer [Governeur, N.Y.], frame 346, roll 12; survey of Annie E. Sands [Islip, N.Y.], frame 486, roll 12, all in STF Series; Massachusetts Board of Education, *Fifty-Ninth Annual Report of the Board of Education . . .* [1894-95] (Boston: Wright and Potter, 1896), 145-48; "Discussion," *Bulletin of the American Academy of Medicine* 7 (Feb. 1906): 464-67.

31. *Fourteenth Annual Report of the Woman's Christian Temperance Union of Maryland* (Baltimore: Jas. M. Cumming, 1888), 24; *Mail and Express Illustrated Saturday Magazine*, 15 March 1902; *Minutes of the Thirteenth Annual Convention of the Minnesota WCTU*, 80. See also "Practical Workings of the Ainsworth School Physiology Law in New York State," *School Physiology Journal* 6 (Oct. 1896): 21; Sarah S. Low to MHH, 12 May 1902, frames 521-22, roll 12, STF Series; *MSM* 4 (21 Feb. 1884): 45; *Report of the Superintendent of Public Instruction of Washington Territory* (Olympia: Thomas H. Cavanaugh, 1887), 64; *Biennial Report of Iowa* [1886-87], 32.

32. David Tyack, *The One Best System: A History of American Urban Education* (Cambridge: Harvard University Press, 1974), 97-104; Murphy, *Blackboard Unions*, 24-34; Fuller, *Old Country School*, 216; Alice Yageric, "Instruction in the Effects of Alcohol and Narcotics in New York State Public Schools, 1884-1933" (Ph.D. diss., State University of New York at Buffalo, 1995), 109-10; "Temperance Instruction in

Philadelphia," *Union Signal,* 23 May 1889; report by Anna Holyoke Howard in *Union Signal,* 21 Jan. 1886.

33. Survey of E. W. Tooker [Riverhead, N.Y.], frame 503, roll 12, STF Series; *Eleventh Annual Report of the Woman's Christian Temperance Union of Illinois* (Chicago: Jameson and Morse, 1884), 90; Leo G. Schussman, "Scientific Temperance Instruction and the School Master," *Scientific Temperance Journal* 19 (April 1910): 115-17; survey of Caroline R. Davis [Sloanesville, N.Y.], frame 418, roll 12, STF Series.

34. Mary E. Young to MHH, 26 May 1902, frames 187-88, roll 12, STF Series; *Ohio Educational Monthly* 49 (May 1900): 221-22; "Report of the Committee on the Teaching of Hygiene," 11; "Opposed to the New Laws," *New York Times,* 3 July 1895; "Temperance Teaching," *Teachers' Institute* [N.Y.] (Oct. 1900), frame 469, roll 17, STF Series. See also *Protest against the Bill Unnecessarily Increasing Physiology and Hygiene with Special Reference to the Nature of Alcoholic Drinks and Other Narcotics* (n.p., 1895), 116; survey of Ida Russell [Gowanda, N.Y.], frames 634-35, roll 11, STF Series.

35. Edith Wills to F. Gregg, 13 June 1910, box 2, STF Papers; Clifford, "Man/Woman/Teacher," 309-10; *Protest against the Bill,* 88.

36. Tyack, *One Best System;* Magee Pratt, "Compulsory Temperance Education," *Hartford Courant,* 6 April 1901.

37. Hogan, *Class and Reform,* 221; "Teaching Scientific Temperance," *School* 7 (9 Jan. 1896): 154.

38. Untitled editorial, *Union Signal,* 22 March 1883. See also *Wellsville* [Ohio] *Union,* 23 Feb. 1883, frames 113-14, roll 8, STF Series. For other examples of Hunt and her allies defending teacher professionalism in the 1880s, see "Temperance in Schools," *MSM* 5 (18 Sept. 1884): 46; Emma Ward Bumstead, "Our New England Letter," *Union Signal,* 5 Jan. 1888.

39. Untitled editorial, *School Physiology Journal* 3 (October 1894): 1; "Teachers Defy the Law," *Union Signal,* 8 August 1895.

40. JoAnne Brown, *The Definition of a Profession: The Authority of Metaphor in the History of Intelligence Testing, 1890-1930* (Princeton: Princeton University Press, 1992), 48; Susanna M. D. Fry, "Active Opposers of Scientific Temperance Instruction," *Union Signal,* 6 March 1902; S. S. Weatherby, "Kansas: To the Front Again," *Union Signal,* 2 Feb. 1891; untitled editorial, *School Physiology Journal* 8 (Sept. 1898): 14. By the late 1890s, Hunt had joined the chorus of critics demanding sharp restrictions on immigration to the United States. See, e.g., MHH to Frances E. Willard, 4 Oct. 1897, frames 625-27, roll 13, STF Series; MHH, "Sample Lecture" (MS, n.d.; [1898?]), frames 78-79, roll 14, STF Series.

41. "What One Y Union Did for Temperance Teaching," *Union Signal,* 25 June 1891; Frances C. Pierce Smith to MHH, 13 May 1902, frame 536, roll 11, STF Series; MHH to Maggie S. Staat, 13 Nov. 1901, frames 693-94, roll 7, STF Series.

42. On STI curricula at normal schools and teacher institutes, see, e.g., *Minutes of the Ninth Convention of the National Woman's Christian Temperance Union* (Brooklyn: Martin, Carpenter, and Co., 1882), xc; *Annual Report of the Board of Education of the State of Connecticut . . .* (New Haven: Tuttle, Morehouse and Taylor, 1881), 96-97; Clarissa L. Ware, "Temperance Teaching in Wisconsin," *Union Signal,* 1 Oct. 1885; remarks of "Superintendent Ellis" [Rochester], 1885 minutes, box 1, Council of School Superintendents; "Physiology for the Lower Grades," *MSM* 5 (19 March 1885): 529. For examples of WCTU women addressing institutes, see Agnes D. Hays, *The White Ribbon in the Sunflower State* (Topeka: Woman's Christian Temperance Union of Kansas, 1953), 88; N. E. White, "Normal Instruction and Teachers' Institutes," *Eleventh Annual Report of the Woman's Christian Temperance Union of*

Pennsylvania (Pittsburgh: Anderson and Son, 1885), 77-83; N. E. White, "Work among Teachers and Students," *Twelfth Annual Report of the Woman's Christian Temperance Union of Pennsylvania* (Pittsburgh: James McMillan, 1886), 180-83.

43. Leila S. Hutton to MHH, 9 May 1902, frame 726, roll 12, STF Series; Physiology Notebook of Nettie Simpson [1903], Maryland State Normal School Records, Towson State University Archives, Towson, Maryland; Irving Bishop, et al., "Scientific Instruction in the Effects of Stimulants and Narcotics," *Educational Review* 24 (1902): 41-42. Just thirteen years earlier, another Maryland normal student's notebook recorded detailed lectures about alcohol's effect upon the stomach, tissues, cells, and liver. Physiology Notebook of Hilda Crass, 19 Feb. 1889, 28 Feb. 1889, 19 April 1889, 7 May 1889, 8 May 1889, 9 May 1889, 10 May 1889, Maryland State Normal School Records.

44. *School Physiology Journal* 5 (Nov. 1895): 44; *Minutes of the Eighteenth Convention of the National Woman's Christian Temperance Union* (Chicago: Woman's Temperance Publishing Association, 1891), 182; *The Inter-Ocean,* 15 Oct. 1891; William R. Johnson, "Teachers and Teacher Training in the Twentieth Century," in Warren, ed., *American Teachers,* 245-49.

45. MHH to Samuel L. Beiler, 20 Aug. 1894; Beiler to Susanna M. D. Fry, 7 Aug. 1894; Beiler to MHH, 9 Aug. 1894, all in folder AC 5810-1, AU Papers; Richard Wheatley, "The American University," *The University Courier* [American U.] 4 (Dec. 1895): 6-7. See also letter from Mary Hunt, "Physiological Temperance in Public Schools," *New York Evening Post,* 3 June 1895.

46. Beiler to MHH, 9 Aug. 1894; MHH to Beiler, 20 Aug. 1894; Beiler to Fry, 4 Sept. 1894, all in folder AC 5810-1, AU Papers.

47. Beiler to Fry, 4 Sept. 1894; "Agreement between the Board of Counsel of the Temperance Educational Association and the Trustees of the American University, Washington, D.C." (MS, 5 Dec. 1895), both in folder AC 5810-1, AU Papers. See also "A Forward Movement," *School Physiology Journal* 5 (Jan. 1896): 72.

48. Beiler to Fry, 4 Sept. 1894; Fry to Beiler, 12 Sept. 1894; MHH to Beiler, 5 Oct. 1896, all in folder AC 5810-1, AU Papers.

49. Hunt appealed to Carnegie-Mellon University, where Andrew Carnegie told her he "had not time to investigate" her request; to the University of Chicago, where president William Rainey Harper had already established a "Theological School" and begged divisional overload; and finally to the new Simmons College in Boston, which seemed the last best hope for an STI college. Yet here Hunt's 1902 inquiries were quickly squelched by MIT biologist William T. Sedgwick, a friend of Wilbur Atwater and a member of the Simmons board of trustees. MHH to William A. Mowry, 24 Nov. 1901, frames 822-23, roll 7; MHH to Andrew Carnegie, 24 Feb. 1904, frames 230-34, roll 15; Susanna M. D. Fry to MHH, 11 Dec. 1896, frame 432, roll 1; Fredericka C. Babcock to MHH, 2 May 1902, frame 28, roll 3, all in STF Series.

50. Mary H. Hunt, "Temperance Education in the Middle States—No Reaction against It among Teachers," *School Physiology Journal* 13 (Jan. 1904): 78-79; "Reply of Representative Temperance People of America to the Proposal of the Central Association of Science and Mathematics Teachers . . ." (MS, n.d. [Feb. 1906?]), frames 60-64, roll 66, STF Series; "Moral Suasion versus Scientific Temperance Instruction," *Union Signal,* 26 April 1906.

51. *Report of the Twenty-Seventh Annual Meeting of the Woman's Christian Temperance Union of the State of New York* (Ithaca: Press of the Ithaca Journal, 1900), 168-69; Murphy, *Blackboard Unions,* 12; Michael W. Sedlak, "'Let Us Go and Buy a School Master': Historical Perspectives on the Hiring of Teachers in the United States, 1750-1980," in Warren, ed., *American Teachers,* 266.

6. "Do-Everything"

1. Katharine Stevenson, "Justice within Our Gates," *Union Signal*, 7 August 1902, frame 803, roll 15, Temperance and Prohibition Papers, Ohio Historical Society (joint Ohio Historical Society—Michigan Historical Collections), Scientific Temperance Federation Series [hereafter STF Series]; Katharine Louise Stevenson, "Compromise for the Sake of Harmony," *Union Signal*, 15 May 1902, frame 798, roll 15, STF Series.

2. For a full account of this struggle, see Chapter Two of this book.

3. See, e.g., Ruth Bordin, *Woman and Temperance: The Quest for Power and Liberty, 1873-1900* (Philadelphia: Temple University Press, 1981), esp. 152; Barbara Leslie Epstein, *The Politics of Domesticity: Women, Evangelism, and Temperance in Nineteenth-Century America* (Middletown, Conn.: Wesleyan University Press, 1981).

4. On this "pipeline" mode of leadership succession within the WCTU, see Joseph Gusfield, "The Problem of Generations in an Organizational Structure," *American Sociological Review* 35 (1957): 326-27.

5. Clipping, *Washington Post*, 21 Feb. 1891, Willard Scrapbook no. 54, National WCTU Scrapbooks, Frances E. Willard Memorial Library, Evanston, Illinois; Katharine Stevenson to MHH, 19 Sept. 1895, frames 352-53, roll 1, STF Series; Stevenson to Frances Willard, 15 Dec. 1896, box 10, Wilbur O. Atwater Papers, Olin Library, Wesleyan University, Middletown, Connecticut [hereafter Atwater Papers].

6. Stevenson to Frances Willard, 15 Dec. 1896, box 10, Atwater Papers; Stevenson to Atwater, 16 July 1897, box 10, Atwater Papers.

7. Atwater to Stevenson, 25 Oct. 1897, box 10, Atwater Papers. See also Stevenson to Atwater, 16 July 1897, box 10, Atwater Papers.

8. Stevenson to Atwater, 11 May 1898, box 4; Stevenson to Atwater, 26 Nov. 1898, box 4; Atwater to Stevenson, 16 March 1897, box 8; Stevenson to Atwater, 26 March 1897, box 10; Stevenson to Atwater, 2 July 1897, box 10, all in Atwater Papers.

9. Massachusetts Board of Education, *Sixty-Third Annual Report of the Board of Education* . . . [1898-99] (Boston: Wright and Potter, 1899), 543-49; *Documents Printed by the Order of the Senate of the Commonwealth of Massachusetts*, no. 282 (n.p., 1899); *Temperance Teaching in the Schools* (n.p., 1899), frame 416, roll 14, STF Series; "Scientific Temperance," *Journal of Education*, 23 March 1899, frame 419, roll 14, STF Series; Henry P. Bowditch and Clifton F. Hodge, "Report on the Present Instruction on the Physiological Action of Alcohol," in John S. Billings, ed., *Physiological Aspects of the Liquor Problem*, 2 vols. (Boston: Houghton Mifflin, 1903), 1:128-34; Bowditch to Atwater, 24 Feb. 1899; Bowditch to Atwater, 3 April 1899, both in box 4, Atwater Papers.

10. "The Massachusetts Central Committee's Report," *School Physiology Journal* 8 (April 1899), extra ed.; Bowditch to Atwater, 20 Jan. 1900, box 5, Atwater Papers; Stevenson to Atwater, 29 Nov. 1899, box 4, Atwater Papers.

11. Frances Willard, *Occupations for Women* (New York: Success Co. 1897), 265-66; Ruth Bordin, *Frances Willard: A Biography* (Chapel Hill: University of North Carolina Press, 1986), 126, 131; Bordin, *Woman and Temperance*, 102-3; Ellen M. Henrotin, "The Co-operation of Woman's Clubs in the Public Schools," *Journal of Proceedings and Addresses of the National Education Association* 36 (1897): 75. For other examples of Willard embracing child-study theory, see Willard, *Occupations for Women*, 267-68; Judith B. Erikson, "Making King Alcohol Tremble: The Juvenile Work of the Woman's Christian Temperance Union, 1874-1900," *Journal of Drug Education* 18 (1988): 341.

12. George W. Fitz to Bowditch, 26 Jan. 1900, frame 166, roll 11; Fitz to

Bowditch, 26 Jan. 1900, frame 167, roll 11; Mary H. Hunt, "Getting Together," *Union Signal,* 20 Jan. 1900, frame 78, roll 17; A. H. Plumb and Joseph Cook to Mary G. Stuckenberg, 14 Dec. 1899, frame 14, roll 17, all in STF Series. See also MHH to Stevenson, 24 Nov. 1899, frames 885-87, roll 6; MHH to Stevenson, 18 Jan. 1900, frames 346-47, roll 17, both in STF Series.

13. "A Plain Statement of Facts" (MS, n.d. [1901?]), frames 539-49, roll 15, STF Series; "News From the Field—Massachusetts," *Union Signal,* 9 Feb. 1899; *Report of the National Woman's Christian Temperance Union. Twenty-Eighth Meeting* (Chicago: Woman's Temperance Publishing Association, 1901), 212. See also "Purington, Mary Louise," *Standard Encyclopedia of the Alcohol Problem,* 6 vols. (Westerville, Ohio: American Issue, 1925-30), 5:2228.

14. "A Plain Statement of Facts"; Mary Hunt, "Preliminary Statement (MS, n.d. [1900?]), frame 639, roll 15, STF Series; "The Minority Report" (MS, n.d. [1900?]), frame 585, roll 15, STF Series; *Union Signal,* 17 July 1902. See also "The Real Issue," *School Physiology Journal* 12 (Dec. 1902): 57, 65. The other signatories to Hunt's minority report were Roxbury minister Albert Plumb and Jessie Forsyth, a longtime official in the Independent Order of Good Templars. See David Fahey, ed., *The Collected Writings of Jessie Forsyth, 1847-1937: The Good Templars and Temperance Reform on Three Continents* (Lewiston, N.Y.: Edwin Mellen Press, 1988).

15. Hunt, "Getting Together"; MHH to Stevenson, 18 Jan. 1900, frames 346-47, roll 17, STF Series.

16. Katharine Stevenson to "Dear Sister State-President," 11 March 1900, frame 650, roll 16, STF Series.

17. MHH to Lillian Stevens, 23 Aug. 1901, frame 774; MHH to Stevens, 14 Sept. 1901, frames 777-81, both in roll 14, STF Series; MHH to Katharine Stevenson, 18 Jan. 1900, frames 346-47, roll 17, STF Series; L. M. N. Stevens, "The Maine Law All Right," *Union Signal,* 11 June 1885. For more on Stevens's doctrine of "states' rights," see "Temperance Lessons," *Boston Herald,* 22 Oct. 1902, frame 698, roll 16, STF Series.

18. Katharine Stevenson, "Compromise for the Sake of Harmony," *Union Signal,* 15 May 1902; idem, "Justice within Our Gates"; Cornelia B. Forbes to MHH, 12 Feb. 1901, frames 822-25, roll 1, STF Series; "Memorandum of Conversation," 21 Feb. 1901, box 10, Atwater Papers; Atwater to Bowditch, 27 Feb. 1901, frames 762-63, roll 10, STF Series; "Reply of Mr. W. B. Ferguson, superintendent of schools, Middletown, to objections made against the substitute educational temperance bill by Mrs. Mary H. Hunt and others" (MS, 1901), box 10, Atwater Papers.

19. Geraldine Joncich Clifford, "Man/Woman/Teacher: Gender, Family, and Career in American Educational History," in *American Teachers: Histories of a Profession,* ed. Donald Warren (New York: Macmillan, 1989), 315; Stevenson, "Justice Within Our Gates"; Anna Pollard quoted in W. B. Ferguson, "Temperance Teaching and Recent Legislation in Connecticut," *Educational Review* 23 (March 1902): 242.

20. See, e.g., Clara Hoffman to Mary Lovell, 8 Jan. 1902, frames 206-10, roll 15, STF Series.

21. Stevenson, "Compromise for the Sake of Harmony."

22. MHH to Cora Stoddard, 12 Aug. 1902, frames 102-4, roll 1, STF Series; Mary Hunt, "The Present Need and How to Meet It," *Union Signal,* 24 April 1901; *Twenty-Ninth Annual Report of the Woman's Christian Temperance Union of Michigan* (Bay City: John P. Lambert, 1903), 108-10. See also Mary Hunt, "Shall Yours Be a Banner State?" *North Carolina White Ribbon* (Jan. 1902), frame 479, roll 17, STF Series.

23. MHH to Mattie G. Shook, 9 Aug. 1901, frame 220, roll 16, STF Series. For other examples of the "ammunition" metaphor, see Hunt circular to New Jersey

county WCTU presidents, 27 Sept. 1901, frame 108, roll 7, STF Series; Lucy Edmunds to Mary Lovell, 3 Feb. 1890, frames 570-71, roll 16, STF Series.

24. M. McDuffie to Elizabeth Rehm, frames 873-74, roll 5; MHH to S. S. Rodocker, 14 Nov. 1899, frames 752-54, roll 6, both in STF Series.

25. MHH to Margaret C. Muhns, 20 Sept. 1901, frames 19-20, roll 7, STF Series; Lytie P. Davies, "To County and Local Superintendents," n.d., box 5, Scientific Temperance Federation Papers, New York Public Library, New York, New York.

26. See, e.g., MHH to Winfield S. Hall, 14 Sept. 1899, roll 6, frames 78-79, STF Series.

27. *Twenty-Fifth Annual Meeting of the Woman's Christian Temperance Union of the State of Kansas* (Newton: Evening Kansan, 1903), 89; Paul Boyer, *Urban Masses and Moral Order in America, 1820-1920* (Cambridge, Mass.: Harvard University Press, 1978), 190, 278; Robert H. Wiebe, *The Search For Order, 1877-1920* (New York: Hill and Wang, 1967), 145.

28. N. L. Caminade, "N.J.: Redeemed," *Union Signal*, 10 May 1894; "The Black Spot Removed," *New York Tribune*, n.d. [1894], frame 421, roll 17, STF Series; *Report of the Twenty-Sixth Meeting of the New York WCTU* (Ithaca, N.Y.: Press of the Ithaca Journal, 1899), 176; *Eighth Annual Meeting of the Woman's Christian Temperance Union of Iowa* (Des Moines: Iowa Commonwealth, 1897), 103.

29. Cora Stoddard to Mary Lovell, 29 May 1906, frames 418-26, roll 13, STF Series; M. Emma Smith to E. O. Taylor, 10 May 1906, frames 532-35, roll 3, STF Series; Frances Whitaker, "A History of the Ohio Woman's Christian Temperance Union, 1874-1920" (Ph.D. diss., Ohio State University, 1971), 307; Mary H. Hunt, "Present Issues in Scientific Temperance Instruction," enclosed with Cora Stoddard to E. O. Taylor, 1 Aug. 1902, frame 792, roll 15, STF Series.

30. E. O. Taylor to MHH, 10 Sept. 1902, frames 79-80, roll 3; Taylor to Cora F. Stoddard, 29 July 1902, frames 52-58, roll 3; MHH to Stoddard, 11 Aug. 1902, frames 99-100, roll 1, all in STF Series.

31. "Temperance Lessons," *Boston Herald*, 22 Oct. 1902, frame 698, roll 16; MHH to Stevens, 20 July 1903, frames 215-16, roll 15, both in STF Series. See also *Report of the National Woman's Christian Temperance Union. Twenty-Ninth Meeting* (Chicago: Woman's Temperance Publishing Association, 1902), 88-89.

32. *Report of the Twenty-Sixth Annual Meeting of the WCTU of New York*, 176. Illinois, Massachusetts, and South Dakota also saw repeal bills fizzle. Mary L. Brumbach to MHH, 2 Feb. 1899, frames 665-69, roll 2, STF Series; Massachusetts Board of Education, *Sixty-Third Annual Report of the Board of Education . . .* [1898-99] (Boston: Wright and Potter, 1900), 543-49; *Tenth Annual Meeting of the Woman's Christian Temperance Union of Iowa* (Lyons: W. B. Farver, 1899), 86-87.

33. Ferguson, "Temperance Teaching and Recent Legislation," 233. Between 1901 and 1904, Arkansas, Illinois, Louisiana, and Wisconsin defeated bills to weaken their respective STI laws; New York and Ohio, meanwhile, fought off attempts at outright repeal. Susanna M. D. Fry, "Corresponding Secretary's Notes," *Union Signal*, 9 June 1904; MHH to Laura N. Steele, 13 Aug. 1901, frame 767, roll 14, STF Series; Othel Lovell, "Crucial Issues in Louisiana Public Education as Reflected in Selected Louisiana Newspapers, 1884-1956" (Ph.D. diss, Louisiana State University, 1961), 184-85; Lucy Gould Baldwin, "Louisiana: Scientific Temperance Instruction," *Union Signal*, 28 Aug. 1902; Hunt, "Getting Together"; *Report of the National WCTU. Twenty-Ninth Meeting*, 347; "Superintendents," *Union Signal*, 29 Dec. 1904; "Dark-Lantern Diabolism," *American Issue*, 11 March 1904, frame 492, roll 17, STF Series.

34. *Minutes of the Thirtieth Annual Meeting of the Woman's Christian Temperance Union of the State of Minnesota* (Minneapolis: Thurston and Gould, 1906), 34, 77-78;

Report of the Sixteenth Annual Meeting of the Woman's Christian Temperance Union of North Dakota (Bismarck: *Bismarck Tribune*, 1905), 32.

35. Philip J. Pauly, "The Struggle for Ignorance about Alcohol: American Physiologists, Wilbur Olin Atwater, and the Woman's Christian Temperance Union," *Bulletin of the History of Medicine* 64 (1990): 388–89; "Resolutions Presented to the Members of the Convention," *Union Signal*, 30 Nov. 1905; "Refusal to Co-Operate with the Committee of Fifty," *School Physiology Journal* 15 (Jan. 1906): 76–77.

36. Elizabeth Preston Anderson to MHH, 2 Feb. 1906, frame 770; MHH to Anderson, 28 Feb. 1906, frames 775-78, both in roll 17, STF Series.

7. "LOCAL INFLUENCE" ON THE WANE

1. William H. Allen, "A Broader Motive For School Hygiene," *Atlantic Monthly* 101 (June 1908): 824, 829.

2. Allen, "Broader Motive," 825; "The Scientific Temperance Federation," *School Physiology Journal* 17 (Jan. 1908): 79-80a.

3. The standard history of school hygiene and health education is still Richard K. Means, *A History of Health Education in the United States* (Philadelphia: Lea and Febiger), 1962. See also William J. Reese, *Power and the Promise of School Reform: Grassroots Movements during the Progressive Era* (Boston: Routledge and Kegan Paul, 1986), ch. 8; John Duffy, *The Sanitarians: A History of American Public Health* (Urbana: University of Illinois Press, 1990), 181-83, 210-13; JoAnne Brown, *The Definition of a Profession: The Authority of Metaphor in the History of Intelligence Testing, 1890-1930* (Princeton: Princeton University Press, 1992), 51-61.

4. The phrase "social bonds" comes from Daniel T. Rodgers, "In Search of Progressivism," *Reviews in American History* 10 (1982): 123. For discussions of this theme, see Thomas L. Haskell, *The Emergence of Professional Social Science: The American Social Science Association and the Nineteenth-Century Crisis of Authority* (Urbana: University of Illinois Press, 1977); Wilfred M. McClay, *The Masterless: Self and Society in Modern America* (Chapel Hill: University of North Carolina Press, 1994); R. Jackson Wilson, *In Quest of Community: Social Philosophy in the United States, 1860-1920* (New York: Wiley, 1968). On Progressives and alcohol, the standard text is still James H. Timberlake, *Prohibition and the Progressive Movement, 1900-1920* (Cambridge: Harvard University Press, 1963). I have also relied heavily upon K. Austin Kerr, *Organized for Prohibition: A New History of the Anti-Saloon League* (New Haven: Yale University Press, 1985).

5. Means, *History of Health Education*, ch. 4.

6. Maurice Bigelow, "The Place of Health Education in the Programs of Schools and Colleges," *Report of Conference on Health Education and the Preparation of Teachers* (New York: Child Health Organization, 1922), 59; "Report of the Committee on the Teaching of Hygiene in Public Schools," *Bulletin of the American Academy of Medicine* 7 (1905): 11, 14. The term "careerist ambitions" comes from Thomas Haskell, "Power to the Experts," *New York Review of Books*, 17 Oct. 1977, p. 33.

7. Michael McGerr, "Political Style and Women's Power, 1830-1930," *Journal of American History* 77 (Dec. 1990): 884.

8. "The Hygiene Examination of the Future," *School Bulletin* 22 (Oct. 1895): 28. For another satire linking STI to school hygiene, see *Cincinnati Enquirer*, 2 March 1888.

9. James C. Whorton, *Crusaders for Fitness: The History of American Health Reformers* (Princeton: Princeton University Press, 1982), ch. 2; Ronald G. Walters, *American Reformers, 1815-1860* (New York: Hill and Wang, 1978), ch. 7; Horace Mann, *Life and Works of Horace Mann*, 5 vols. (Boston: Lee and Shepard, 1891)

3:439-41; Horace Mann, *The Study of Physiology in the Schools* (New York: J. W. Schermerhorn, 1869), 4.

10. Mary H. Hunt, *An Epoch of the Nineteenth Century: An Outline of the Work for Scientific Temperance Instruction in the Public Schools of the United States* (Boston: P. H. Foster, 1897), 44; "Standard Lecture" (MS, n.d.), frames 30-31, roll 14, Temperance and Prohibition Papers, Ohio Historical Society (joint Ohio Historical Society—Michigan Historical Collections), Scientific Temperance Federation Series [hereafter STF Series]; "The Doctors and the Scientific Temperance Education Bill" (MS, n.d. [1886?]), frame 918, roll 9, STF Series; Mary H. Hunt, *A History of the First Decade of the Department of Scientific Temperance Instruction . . .* , 2nd ed. (Boston: George E. Crosby, 1891), 8.

11. "Report of the Committee on the Teaching of Hygiene in Public Schools," *Bulletin of the American Academy of Medicine* 7 (Feb. 1906): 419; Cora Stoddard to "Dear Sir," 28 May 1904, frames 31-33, roll 9, STF Series. For other examples of sanitarians' efforts to widen school health curricula during the Hunt years, see, e.g., George G. Groff, "Thoughts on School Hygiene," *Sixth Annual Report of the State Board of Health . . .* [Pennsylvania] (Harrisburg: Edwin K. Meyers, 1891), 523-26; C. O. Probst, "Instruction in Hygiene in Schools and Colleges," *Public Health Papers and Reports* [American Public Health Association] 20 (1894): 251-58; George A. Soper, "Hygiene as a Factor in Education," *Educational Review* 24 (Nov. 1902): 391-97; Benjamin Lee, "Report of the Committee on the Teaching of Hygiene and the Conferring of a Degree in Public Health," *Public Health Papers and Reports* 29 (1903): 124-27; "Report of Committee to Examine School Text-Books," *Pennsylvania Medical Journal* 8 (1904-5): 39-41.

12. Hunt, *History of the First Decade*, 41; Hunt, *Epoch of the Nineteenth Century*, 34; "School Instruction in the Effects of Stimulants and Narcotics," *Educational Review* 24 (1902): 44. See also James E. Peabody, "Physiology in the High School," *New York Teachers' Magazine* 1 (Feb. 1899), 164; Groff, "Thoughts on School Hygiene," 523; Lee, "Report of the Committee on the Teaching of Hygiene," 124-25; Probst, "Instruction in Hygiene in Schools and Colleges," 251-52.

13. George G. Groff, "Methods of Teaching Physiology and Hygiene," *Pennsylvania School Journal* 35 (Feb. 1887): 316; George G. Groff, "Physiology Versus Hygiene in Our Public Schools," *Bulletin of the American Academy of Medicine* 6 (Aug. 1903), frames 24-29, roll 9, STF Series; Lee, "Report of the Committee," 124. See also George G. Groff, "Science Teaching in Secondary Schools," *Science* 22 (20 Oct. 1893), frames 432-33, roll 9, STF Series; George G. Groff, "On Teaching the Principles of Hygiene to the Young," *Public Health Papers and Reports* 20 (1894): 263-67.

14. Charles R. Skinner to Frederick W. Holls, 8 Oct. 1895, box 7, Frederick W. Holls Papers, Butler Library, Columbia University; Soper, "Hygiene as a Factor in Education," 397. See also Thomas D. Wood, "The Scope of School Hygiene in Modern Education," *Teachers College Record* 6 (March 1905): 1-13.

15. Robyn Muncy, *Building a Female Dominion in American Reform, 1890-1935* (New York: Oxford University Press, 1991), 30, 60.

16. For all their talk about influencing the curriculum, sanitarians rarely stepped foot into a school. Like other reformers, they presumed that a properly conceived reform could operate by "remote control." See Paul Boyer, *Urban Masses and Moral Order in America, 1820-1920* (Cambridge: Harvard University Press, 1978), 277-78. But this vision also presumed a passive, acquiescent public—something very different from Hunt and her local army of WCTU women.

17. Cora Stoddard to Mary Lovell, 18 May 1904, frames 221-22, roll 17; Stoddard to "Dear Sir," frames 31-33, roll 9; "Reply of Representative Temperance

People of America to the Proposal of the Central Association of Science and Mathematics Teachers to Substitute Moral Suasion for Scientific Temperance Instruction in Our Public Schools" (MS, 1906), frames 60-64, roll 66, all in STF Series. For more on the American Academy of Medicine and its school hygiene committee, see Steven J. Peitzman, "Forgotten Reformers: The American Academy of Medicine," *Bulletin of the History of Medicine* 58 (1984): 516-28; Jonathan Zimmerman, "'The Queen of the Lobby': Mary H. Hunt, Scientific Temperance, and the Dilemma of Democratic Education in America" (Ph.D. diss., Johns Hopkins University, 1993), 320-23. On the teachers' proposal, see also "Matters on Physiology and Hygiene," *School Science* 5 (1905):111-13; "Report of the committee . . . ," ibid., 5: 203-7; "Moral Suasion versus Scientific Temperance Instruction," *Union Signal,* 26 April 1906.

18. "Reply of Representative Temperance People," frame 64; Roy Blue, "The New Law," *Indiana School Journal* 15 (September 1895): 567-68; Ida F. Dickerson, "'The New Law' Again," *Indiana School Journal* 15 (Oct. 1895): 640-41; Massachusetts Board of Education, *Sixty-Third Annual Report of the Board of Education . . .* [1898-99] (Boston: Wright and Potter, 1900), 575.

19. "Informal Gathering" (MS, May 1906), frames 32-36, roll 39, STF Series. For biographical background on Taylor, see Cora Stoddard, "A Jubilee of Public Service," enclosed with Stoddard to "Editor of *Patriot,*" 23 Sept. 1913, frames 395-96, roll 24, STF Series.

20. "Informal Gathering," frames 52-53, 32-34. See also "Biography of Emma L. Transeau, ne Benedict . . ." (MS, 1921), frames 349-66, roll 15, STF Series.

21. Susanna M. D. Fry to A. H. Plumb, 29 Aug. 1906, box 2, Scientific Temperance Federation Papers, New York Public Library, New York, New York [hereafter STF Papers]; "Scientific Temperance Association" (MS, 23 June 1906), frames 780-89, roll 17, STF Series. See also "Scientific Temperance Association" (MS, n.d. [1906]), frames 800-801, roll 17, STF Series.

22. Richard L. McCormick, "The Discovery that Business Corrupts Politics: A Reappraisal of the Origins of Progressivism," *American Historical Review* 86 (April 1981): 259-64; "Minutes of meeting at National WCTU convention" (MS, 29 Oct. 1906), frame 587, roll 27, STF Series; Cora Stoddard to Mary Lovell, 24 Jan. 1907, frames 816-17, roll 17, STF Series; E. O. Taylor to Minnie E. Neal, 2 Aug. 1907, frame 668, roll 45, STF Series; MHH to Frances Willard, 23 June 1892, frames 281-82, roll 15, STF Series; MHH to Lillian Stevens, 20 July 1903, frame 226, roll 15, STF Series; *Report of the National Woman's Christian Temperance Union. Thirty-Third Convention* (Chicago: Woman's Temperance Publishing Association, 1906), 56-57. Scientific Temperance Association members included Crothers, a leading voice on the STF; schoolman William Mowry and minister Albert H. Plumb, both longtime stalwarts of Hunt's "Advisory Board"; and New Hampshire senator and temperance activist Henry Blair, who served as president of the STA for several years.

23. Scientific Temperance Association, "Statement" (MS, 1906), frames 253-57, roll 15, STF Series. For more detail on the textbook trade, see Chapter Two of this book.

24. Scientific Temperance Association, "Statement," frames 253-57, roll 15, STF Series; MHH to Frances Willard, 23 June 1892, frames 281-82, roll 15, STF Series; D. C. Heath to Wilbur Atwater, 30 Jan. 1901, box 5, Wilbur O. Atwater Papers, Olin Library, Wesleyan University, Middletown, Connecticut [hereafter Atwater Papers]; M. F. Lovell, "The Work of a Remarkable Woman and Its Results" (MS, April 1927), frame 174, roll 66, STF Series; "Minutes of meeting at National WCTU convention," frame 564; "Scientific Temperance Association" (MS, n.d. [1906]), frames 800-801, roll 17, STF Series.

25. "Minutes of meeting at National WCTU convention," frames 570, 576, 575; *Report of the Thirty-Third Convention,* 58.

26. Elizabeth Preston Anderson to MHH, 2 Feb. 1906, frame 770, roll 17, STF Series; Jane A. Stewart, "A Leader in Scientific Temperance Instruction," *Union Signal,* 8 Sept. 1910; "Wisconsin State Convention," *Union Signal,* 24 Oct. 1907; Circular (n.p., 11 Dec. 1906), frame 818, roll 17, STF Series.

27. "Taken from 'Woman's Temperance Work,' 'Official Organ' of 'The Woman's Christian Temperance Union,' State of New York" (MS, n.d. [1906]), box 2, STF Papers; Circular (n.p., 11 Dec. 1906), frame 818, roll 17, STF Series; Louie Winfield Webb, "A Pedagogical Estimate of Scientific Temperance Instruction" (M.A. thesis, University of Chicago, 1913), 49-50; *Report of the National Woman's Christian Temperance Union. Thirty-Fourth Convention* (Chicago: Woman's Temperance Publishing Association, 1907), 234-39.

28. "Gratifying Progress in Work of Scientific Temperance Education," *Union Signal,* 16 May 1907; "Mrs. Edith Smith Davis in New England," *Union Signal,* 4 Feb. 1909; "Here and There," *Union Signal,* 7 Aug. 1911; "Temperance Teaching in the Colleges," *Union Signal,* 6 Aug. 1908; E. S. Davis, "Scientific Temperance Instruction," *Union Signal,* 19 Dec. 1907. See also Norton Mezvinsky, "The White-Ribbon Reform" (Ph.D. diss., University of Wisconsin, 1959), 188; Webb, "A Pedagogical Estimate of STI," 49-50; Stewart, "A Leader in Scientific Temperance Instruction." For Jordan's earlier criticisms of STI, see David Starr Jordan, "Scientific Temperance," *Popular Science Monthly* 48 (1896): 343-54.

29. Minutes of "Special Meeting" of the Scientific Temperance Federation (MS, 24 April 1907), frames 108-9, roll 39; Stoddard to Mary Lovell, 1 Jan. 1907, frames 816-17, roll 17, both in STF Series.

30. Stoddard to Lovell, 24 Jan. 1907, frame 824; Stoddard to Lovell, 8 April 1907, frames 833-35; Stoddard to Lovell, 24 May 1907, frames 839-41, all in roll 17, STF Series.

31. E. O. Taylor to Minnie E. Neal, 2 Aug. 1907, frame 668, roll 45, STF Series; Mrs. J. Frank Smith to Stoddard, 15 Feb. 1909; Smith to Edith Wills, 1 Feb. 1910, both in box 2, STF Papers.

32. As late as 1913, for example, a Quebec WCTU woman asked Stoddard to send her "Mrs. Edith Smith Davis posters" about STI. "I have never seen them," the woman added, "but I am told they are very good and if so we must have them." Mrs. John Scott to Stoddard, 24 July 1913, box 3, STF Papers. For other examples of Canadian correspondence with Stoddard, see L. C. McKinney to Stoddard, 2 Jan. 1908, box 2; F. S. Spence to Stoddard, 27 Oct. 1908, box 2; Ada L. Powers to Stoddard, 14 July 1912, box 3, all in STF Papers.

33. "New Plans," *Scientific Temperance Journal* 19 (Nov. 1909): 7; Stoddard to Wilbur F. Crafts, 22 Oct. 1909, frames 3-9, roll 24, STF Series.

34. "Minutes of meeting at National WCTU convention," frame 576; Stoddard to Marie Brehm, 16 Oct. 1909, frames 1-2, roll 24, STF Series. See also Cora Stoddard, "The London Congress as Seen in Its Personnel," *Scientific Temperance Journal* 19 (Oct. 1909): 22-24.

35. "Introduction of Physiology and Hygiene," *School and Home Education* 28 (April 1909): 278-79; "The Teaching of Physiology," *School Hygiene* 1 (Oct. 1908): 34-35; Allen, "A Broader Motive for School Hygiene"; Means, *History of Health Education,* 85-94, 130-33.

36. Edith Lindsay, "Origins and Development of the School Health Movement in the United States" (Ed.D. diss., Stanford University, 1944), 233; Jesse Knowlton Flanders, *Legislative Control of the Elementary Curriculum* (New York: Teachers Col-

lege, 1925), 82, 66-67. For an overview of health education measures during this era, see Flanders, *Legislative Control,* ch. 3; for a case study, see Corinda Waters, "A Century of Health Instruction in the Public Schools of Maryland, 1872-1972" (Ph.D. diss., University of Maryland, 1972).

37. Helen Barton, "A Study of the Development of Textbooks in Physiology and Hygiene in the United States" (Ph.D. diss., University of Pittsburgh, 1942), 135-37; Means, *History of Health Education,* 102, 106, 88-89; Waters, "A Century of Health Instruction," 65.

38. Guy M. Whipple, "The Instruction of Teachers in School Hygiene," *Pedagogical Seminary* 17 (March 1910): 45. See also J. Mace Andress, *The Teaching of Hygiene in the Grades* (Boston: Houghton Mifflin, 1918), 24-25; Sally Lucas Jean, "Health Education," *Journal of Addresses and Proceedings of the National Education Association* [hereafter *NEA Proceedings*] 58 (1920): 319.

39. "Physiology Next," *The Teacher* [Philadelphia] 15 (Jan. 1911): 356; William A. Link, "Privies, Progressivism, and Public Schools: Health Reform and Education in the Rural South, 1909-1920," *Journal of Southern History* 54 (1988): 641-42; Edith Smith Davis, "Aids in Teaching Temperance in Public Schools," *Union Signal,* 9 Feb. 1911. See also Charles E. North, "Sanitary Education in Rural Communities," *NEA Proceedings* 50 (1912): 267-73.

40. Allen, "A Broader Motive for School Hygiene," 828; Thomas D. Wood, *Ninth Yearbook of the National Society for the Study of Education* (Chicago: University of Chicago Press, 1910), 66.

41. Rodgers, "In Search of Progressivism," 123; remarks by J. S. Grim, in *Conservation of School Children* (Easton, Penn.: American Academy of Medicine, 1912), 108; "Hygiene and Social Education," *School Physiology Journal* 17 (Feb. 1908): 90-91; I. Edwin Goldwasser, "Pupil Cooperation Indispensable in Enforcing Hygienic Regulation," *Transactions of the Fourth International Congress on School Hygiene* 3: 163-67; G. W. Hunter, "An Experiment on Student Control of School Sanitation and Hygiene," ibid., 283-87. See also Philip J. Pauly, "The Development of High School Biology: New York City, 1900-1925," *Isis* 82 (1991): 675.

42. Louis Nussbaum, "How Should Hygiene Be Taught," in *Conservation of School Children,* 106; Seneca Egbert, "Teaching Hygiene in Public Schools," ibid., 75; Herbert Kliebard, "Keeping Out of Nature's Way: The Rise and Fall of Child-Study as the Basis for the Curriculum, 1880-1905," in *Forging the American Curriculum: Essays in Curriculum History and Theory* (New York: Routledge, 1992), 65. On "meliorism" as an educational value, see Kliebard, *Struggle for the American Curriculum,* 24-27, 180-208; for the contrast with efficiency, see ibid., 20-24, 89-122.

43. Allen, "A Broader Motive for School Hygiene," 825; A. Duncan Yocum, "Hygiene Applied to the Modern School," *Pennsylvania School Journal* 59 (Aug. 1910): 99; Hoyt E. Dearholt, "The School vs. the Tuberculosis Problem," *School Board Journal* 41 (Nov. 1910): 7-8; Means, *History of Health Education,* 115-16; William H. Burnham, "The Problems of Child Hygiene, and the Contribution of Hygiene to Education," *NEA Proceedings* 50 (1912): 1101. See also A. Duncan Yocum, "The Teaching of Hygiene in the Elementary School," *American School Board Journal* 37 (Nov. 1908): 3-4.

44. Means, *History of Health Education,* 120, 118; William H. Burnham, *Great Teachers and Mental Health: A Study of Seven Educational Hygienists* (Freeport, N.Y.: Books for Libraries Press, 1926), 267; Andress, *Teaching of Hygiene in the Grades,* 87-88.

45. "Temperance Physiology to Be Taught," *Union Signal,* 1 June 1911. A small handful of institutions did sponsor classes focusing solely on liquor. Yet these courses

also stressed its social aspects: at the University of California, for example, a fourteen-lecture elective on the "scientific investigation" of alcohol included a single class on physiology. Reviewing the field in 1910, Davis could locate only one course in the entire country devoted strictly to the physiology of drink. Offered by Nebraska State University, it had attracted just six students over the previous two years. "Scientific Temperance Instruction Winning Its Rightful Place," *Union Signal*, 13 June 1912; "An Important Course of Study," *Temperance Educational Quarterly* 5 (Jan. 1914): 1; "The Problem of the Teaching of Temperance," *Temperance Educational Quarterly* 1 (Jan. 1910): 17.

46. David S. Jordan, "Temperance and Society," *NEA Proceedings* 49 (1911): 79. On the "social" emphasis of the prohibition movement during these years, see especially Timberlake, *Prohibition and the Progressive Movement*, and Kerr, *Organized for Prohibition*.

47. Jordan, "Temperance and Society," 79; Edith S. Davis, "Suggestive Scientific Temperance Instruction Program," *Union Signal*, 4 Jan. 1912; Edith Davis, *A Compendium of Temperance Truth . . .* (Evanston, Ill.: National Woman's Christian Temperance Union, 1916); "Scientific Temperance Instruction at the Great Summer School of the South," *Union Signal*, 6 Aug. 1908; "The Summer School of the South: WCTU Representative Gives Scientific Temperance Course for Teachers," *Union Signal*, 12 Aug. 1909. See also Stewart, "A Leader in Scientific Temperance Instruction." On the origins and orientation of the *Temperance Educational Quarterly*, see E. S. Davis, "The Drink Dilemma," *Union Signal*, 21 July 1910; "The Problem of the Teaching of Temperance."

48. *Report of the National Woman's Christian Temperance Union. Thirty-Fifth Meeting* (Chicago: Woman's Temperance Publishing Association, 1908), 251; *Twenty-Ninth Annual Report of the Woman's Christian Temperance Union of the State of Delaware* (Dover: Delawarean Print, 1908), 97–98; "Prize Essay Contests," *Union Signal*, 14 Nov. 1918; *Thirty-Third Annual Report of the Woman's Christian Temperance Union of the State of Delaware* (Smyrna: Smyrna Times Print, 1912), 99. For an example of a winning teacher entry, see Virginia S. Milbourne, "The Best Methods of Teaching Temperance," *Temperance Educational Quarterly* 6 (Jan. 1915): 25–27.

49. Edith Davis, "Scientific Temperance Instruction" *Young Crusader* 26 (Aug. 1913): 4; Frances Whitaker, "A History of the Ohio Woman's Christian Temperance Union, 1874–1920" (Ph.D. diss., Ohio State University, 1971), 413–44; Nelle G. Burger, "How Missouri Secured Frances E. Willard Day in the Public Schools," *Union Signal*, 10 Oct. 1918. See also "Plans and Helps for Observing Temperance Day in the Schools," *Temperance Educational Quarterly* 7 (Jan. 1916): 34–37.

50. "Temperance Day," *Union Signal*, 14 Nov. 1918. See also "Plans and Helps"; Burger, "How Missouri Secured"; Cora Stoddard to Anna Gordon, 8 July 1922, frame 203, roll 19, STF Series.

51. "Shall Temperance Instruction Be Social?" *School Physiology Journal* 17 (May 1908): 137; "The Effective Kind of Knowledge," *School Physiology Journal* 17 (April 1908): 123; "The Scientific Temperance Federation," *School Physiology Journal* 16 (June 1907): 198–99. See also Cora Stoddard, "The Practical Relation of Science to the Twentieth-Century Alcohol Problem," *School Physiology Journal* 18 (June 1909): 145–47; "The Heroic Appeal," *Scientific Temperance Journal* 20 (April 1911): 117; "No Information, No Inspiration," *Scientific Temperance Journal* 20 (March 1911): 103.

52. "Report of the Scientific Temperance Federation" (MS, 1911), frames 520–21, roll 54, STF Series. See also "Report of the Scientific Temperance Federation" (MS, 20 Jan. 1914), frames 573–79, roll 54, STF Series; *Reaching the People Where They Are* (n.p., n.d. [1915]), box 5, STF Papers.

53. "Report of the Scientific Temperance Federation" (MS, 20 Jan. 1914), frames 573-79, roll 54, STF Series; Cora Stoddard, "The Place of Education in the Temperance Reform," *Scientific Temperance Journal* 28 (Summer 1919): 74; Cora Stoddard, "Alcoholism at the National Conference of Charities and Conventions," *National Advocate*, July 1911, frame 188, roll 24, STF Series. For other indicators of STF's increasingly "unscientific" approach, see also Stoddard to Emma Transeau, 18 Dec. 1914, box 4, STF Papers; C. J. Atkinson to *Scientific Temperance Journal*, 21 Nov. 1913, box 5, STF Papers; *Reaching the People Where They Are*; Stoddard to Chester Jerome Ford, n.d., frames 371-72; Scientific Temperance Federation circular, n.p., n.d. [1913?], frame 387, both in roll 24, STF Series.

54. Ginn and Co. to P. B. Davis, 11 April 1907, frame 836, roll 17, STF Series; Stoddard to Alfred L. Manierre, 30 Dec. 1909, frames 12-16, roll 24, STF Series; "New Plans," *Scientific Temperance Journal* 19 (Sept. 1909): 7; Stoddard to Charles Scanlon, 29 May 1913, frames 360-62, roll 24, STF Series. See also "Third Annual Meeting of the Scientific Temperance Federation" (MS, 15 Feb. 1910), frame 123; "Meeting of STF Board of Directors" (MS, 1 Feb. 1910), frames 142-43; "Meeting of STF Finance Committee" (MS, 18 Nov. 1909), frame 141, all in roll 39, STF Series.

55. Stoddard to Charles Scanlon, 29 May 1913, frames 360-62, roll 24, STF Series; "Report of the Scientific Temperance Federation" (MS, 1911), frames 520-21, roll 54, STF Series; Lillian Burt to Stoddard, 13 Feb. 1913, box 3, STF Papers; Stoddard to Charles Scanlon, 18 Dec. 1912, frames 326-28, roll 24, STF Series. For the agreement between the League and the STF, see "Special Meeting of the Scientific Temperance Federation" (MS, 19 June 1913), frames 146-49, roll 39, STF Series; to follow the complex negotiations that preceded it, see Stoddard to Ernest Cherrington, 4 Nov. 1912, frames 322-23; Stoddard to C. A. Vincent, 4 Nov. 1912, frame 324; Stoddard to Cherrington, 12 Dec. 1912, frame 325; Stoddard to John W. Cummings, 29 May 1913, frames 358-59, all in roll 24, STF Series. See also Kerr, *Organized for Prohibition*, 157-58.

56. Jack S. Blocker, *American Temperance Movements: Cycles of Reform* (Boston: Twayne Publishers, 1989), 98; *The Warrior's War against Drink: A Wonderful New Set of Stereopticon Slides* (n.p., n.d. [1917?]), box 5, STF Papers; Stoddard, "Alcoholism at the National Conference." The Federation's emphasis upon pithy slogans and commercials neatly mirrored larger developments in national political campaigns, where a new brand of "advertised politics" predominated. Michael E. McGerr, *The Decline of Popular Politics: The American North, 1865-1928* (New York: Oxford University Press, 1986), ch. 6.

57. "An Appreciation: Edith Smith Davis," *Union Signal*, 10 Jan. 1918; "Mrs. Edith Smith Davis, Popular Scientific Temperance Leader and Lecturer, Called to Her Heavenly Reward," *Union Signal*, 28 March 1918; Stoddard Biography (MS, n.d.), frame 220, roll 54, STF Series; "Scientific Temperance and Ratification," *Union Signal*, 10 Jan. 1918; Wilbur F. Crafts to Stoddard, 20 Jan. 1916, box 4, STF Papers.

58. "Bureau of Scientific Temperance Investigation" (MS, n.d. [1906?]), frames 790-94, roll 17, STF Series; Stoddard to J. C. Amspoker, n.d. [1921?], frames 3-5; Stoddard to Frances P. Parks, 3 March 1921, frames 544-46, both in roll 19, STF Series.

59. Frederick Petersen, "Preventive Medicine and the Reconstruction of the Race," *Scientific Temperance Journal* 27 (Dec. 1918): 219-22; Barton, "A Study of the Development of Textbooks," 135-37; Ralph E. Blount, "Phsiology [*sic*] Topics of Most Worth in the Eyes of the Pupils," *Chicago Schools Journal* 4 (Dec. 1921): 137-38; Elizabeth Middleton to Stoddard, 2 Nov. 1921, frame 302, roll 19, STF Series; Stoddard to Margaret Platt, 4 Nov. 1921, frame 601, roll 19, STF Series.

60. Cora Stoddard, "If We Now Do Our Part" (MS, n.d. [1921?], frame 137, roll 66, STF Series; Stoddard to Anna Gordon, 8 July 1922, frames 201-3, roll 19, STF Series; Cora Stoddard, "Scientific Temperance Instruction in the Schools of the United States," *Scientific Temperance Journal* 29 (Dec. 1920): 215. See also Kerr, *Organized for Prohibition,* 263.

61. Elizabeth Middleton to Emma Transeau, 6 May 1922, frame 387; Middleton circular, n.d., enclosed with Middleton to Stoddard, 8 Dec. 1922, frame 429, both in roll 19, STF Series. See also Stoddard to Gordon, 8 Dec. 1922, frames 224-26, roll 19, STF Series.

62. Cora Stoddard, "Plan of Work for 1923" (MS, 1923), frame 105; Stoddard to "Dear State Superintendent," 1 Sept. 1923, frames 131-32, both in roll 20, STF Series.

63. Mrs. A. Swan Brown to Stoddard, 12 Aug. 1924, frames 500-501, roll 22, STF Series; *Minutes of the Forty-Eighth Convention of the National Woman's Christian Temperance Union* (Chicago: Woman's Temperance Publishing Association, 1921), 60, 98; *Report of the Fiftieth Annual Convention of the Massachusetts Woman's Christian Temperance Union* (n.p., 1923), 98-100.

64. Mrs. Howard Park to Stoddard, 20 Feb. 1923, frames 28-29; Anna E. Muzzey to Stoddard, 21 Jan. 1923, frame 438, both in roll 22, STF Series. See also Daisy B. Parlin to Stoddard, 6 Nov. 1923, frame 812, roll 21; Lesta S. Romine to Stoddard, 19 April 1921, frames 173-75, roll 22; Hattie Livermore to Stoddard, 1 Oct. 1924, frames 259-61, roll 22; Ethel Turner to Stoddard, 17 Aug. 1923, frames 345-46, roll 22, all in STF Series.

65. Circular from Mrs. Byron Johnson, 28 Feb. 1923, frame 764, roll 22; Rosa M. Webb to "Dear County S.T.I. Superintendent," frame 372, roll 22; Elizabeth Middleton to Stoddard, 2 Dec. 1924, frames 999-1000, roll 20, all in STF Series. See also Ida W. Smith to "Dear Comrade," n.d. [1923?], frames 460-61, roll 21, STF Series. In the wake of Prohibition, for example, some states mandated special classes in "Respect for Law." To find room for this course, a miffed Cora Stoddard remarked, teachers often *evaded* the law requiring Scientific Temperance. Myrta E. Lockwood to Stoddard, 2 April 1924, frames 319-20; Stoddard to Lockwood, frames 329-30, both in roll 22, STF Series.

66. Stoddard to Middleton, 8 Dec. 1924, frames 1005-6; Stoddard to Anna Gordon, 13 April 1923, frames 326-27; William B. Owen to Frances P. Parks, 11 Nov. 1923, frame 332, all in roll 20, STF Series. For other descriptions of educators' resistance to STI during these years, see Stoddard to Anna Gordon, 30 April 1924, frames 891-92, roll 20; Stoddard to Rosa M. Webb, 24 March 1924, frame 406, roll 22; Stoddard to "Dear Superintendents," 10 March 1923, frame 116, roll 20, all in STF Series. On the NEA's snub of Scientific Temperance, see also Stoddard to Gordon, 14 April 1923, frame 328; Stoddard to Gordon, 20 April 1923, frame 335; Owen to Gordon, n.d., enclosed with Gordon to Stoddard, 26 April 1923, frames 342-43; Stoddard to Gordon, 30 April 1923, all in roll 20, STF Series.

67. Stoddard to Middleton, 8 Dec. 1924, frames 1005-6; Stoddard to Jeannie Kemp, 14 April 1924, frame 941, both in roll 20, STF Series. See also Stoddard to Margaret Munns, 20 Dec. 1924, frames 110-15, roll 21, STF Series.

68. Historians have barely begun to investigate these popular influences upon the early twentieth-century curriculum. For a start, see David Tyack, et al., *Law and the Shaping of Public Education, 1785-1954* (Madison: University of Wisconsin Press, 1987), ch. 6.

69. For a concurring view, see Tyack, et al., *Law and the Shaping of Public Education,* 161.

70. Stephen Arons, *Compelling Belief: The Culture of American Schooling* (New York: McGraw-Hill, 1983).

EPILOGUE

1. W. B. Ferguson, "Temperance Teaching and Recent Legislation in Connecticut," *Educational Review* 23 (March 1902): 239, 235, 242; idem, "Scientific Temperance in Connecticut," *Journal of Education*, 13 June 1901, frame 712, roll 14, Temperance and Prohibition Papers, Ohio Historical Society (joint Ohio Historical Society—Michigan Historical Collections), Scientific Temperance Federation Series [hereafter STF Series].
2. An important exception is David Tyack, et al., *Law and the Shaping of Public Education, 1785-1954* (Madison: University of Wisconsin Press, 1987), esp. ch. 6. For a brief review of the relevant literature, see my Bibliographical Essay.
3. George J. Lankevich, "The Grand Army of the Republic in New York State, 1865-1898" (Ph.D. diss., Columbia University, 1967), 269-71; W. L. Smith, "To All Exalted Cyclops and Kilgrapps," 9 Sept. 1925, folder 2, box 1, Wayne County Ku Klux Klan Records, Indiana Historical Society, Indianapolis, Indiana; Lloyd Arthur Hunter, "The Sacred South: Postwar Confederates and the Sacralization of Southern Culture" (Ph.D. diss., St. Louis University, 1978), 214; Edwin Johnson to Mrs. Newlin T. Paxson, 14 Dec. 1937, reel 70.22, Committee on Militarism in Education Papers, Swarthmore College Peace Collection, Swarthmore, Pennsylvania; William G. Ross, *Forging New Freedoms: Nativism, Education, and the Constitution, 1917-1927* (Lincoln: University of Nebraska Press, 1994), 53; Edward J. Larson, *Trial and Error: The American Controversy over Creation and Evolution* (New York: Oxford University Press, 1985), 48.
4. John Dewey, "Are the Schools Doing What the People Want Them to Do?" *Educational Review* 21 (May 1901): 460-61.
5. John Dewey, "The Situation as Regards the Course of Study," *Educational Review* 22 (June 1901): 31, 48.
6. John Dewey, "Democracy in Education" [1903], in idem, *Education Today*, ed. Joseph Ratner (New York: G. P. Putnam, 1940), 66; Hilary Putnam, "A Reconsideration of Deweyan Democracy," in *Pragmatism, Law and Society*, ed. Michael Brint and William Weaver (Boulder, Colo.: Westview Press, 1991), 240; Dewey, "Are the Schools Doing What the People Want Them to Do?" 472.
7. MHH to Mary F. Lovell, 8 Jan. 1902, frames 206-10, roll 15; survey of Mary F. Donaldson, n.d. [1902], frame 571, roll 12, both in STF Series.
8. Amy Gutmann, "Democratic Education in Difficult Times," *Teachers College Record* 92 (1990): 7-20; Amy Gutmann, *Democratic Education* (Princeton: Princeton University Press, 1987); Jonathan Zimmerman, "'The Queen of the Lobby': Mary Hunt, Scientific Temperance, and the Dilemma of Democratic Education in America, 1879-1906," *History of Education Quarterly* 32 (1992): 1-30.
9. "Just Say Life Skills," *Time* 148 (11 Nov. 1996): 70; Susan T. Ennett, et al., "How Effective Is Drug Abuse Resistance Education? A Meta-Analysis of Project DARE Outcome Evaluations," *American Journal of Public Health* 84 (Sept. 1994): 1394-1401; Earl Wysong, et al., "Truth *and* DARE: Tracking Drug Education to Graduation and as Symbolic Politics," *Social Problems* 41 (Aug. 1994): 453-54; Earl Wysong and David W. Wright, "A Decade of DARE: Efficacy, Politics and Drug Education," *Sociological Focus* 28 (Aug. 1995): 289; David J. Hanson, *Alcohol Education: What We Must Do* (Westport, Conn.: Praeger, 1996), 103, 105.
10. Lloyd P. Jorgenson, *The State and the Non-Public School, 1825-1925* (Columbia: University of Missouri Press, 1987), 76-83; Ross, *Forging New Freedoms*, 45-46;

Howard K. Beale, *Are American Teachers Free? An Analysis of Restraints upon the Freedom of Teaching in American Schools* (New York: Scribner's, 1936), ch. 10.

11. Wysong and Wright, "A Decade of DARE," 301-4. See also Dennis Cauchon, "Studies Find Drug Program Not Effective," *USA Today*, 11 Oct. 1993; idem., "Study Critical of DARE Rejected," *USA Today*, 4 Oct. 1994.

12. Philip J. Pauly, personal correspondence in the possession of the author, 18 Dec. 1990.

BIBLIOGRAPHICAL ESSAY

"So the work goes on," wrote an aide to Mary Hanchett Hunt in January 1887, surveying the achievements of Scientific Temperance Instruction during the previous year. "The historian of the future will find his chief embarrassment in the abundance of material at his command."* She was right. Mandatory in every state and in tens of thousands of school districts, STI generated an enormous stream of books, pamphlets, articles, and personal correspondence. Much of it is preserved in the Scientific Temperance Federation Series of the Temperance and Prohibition Papers, Ohio Historical Society (joint Ohio Historical Society—Michigan Historical Collections), which are catalogued and described in *Guide to the Microfilm Edition of Temperance and Prohibition Papers,* ed. Randall C. Jimerson, et al. (Ann Arbor: University of Michigan Press, 1977). The series features material that an aide collected for an unpublished biography of Mary Hunt, Hunt's voluminous general correspondence, and rich files on her struggle with the Committee of Fifty, her relations with textbook publishers, and a 1902 survey of New York science teachers about STI. It also contains the correspondence of Hunt aide Cora Frances Stoddard, who fell out with the Woman's Christian Temperance Union after Hunt's 1906 death but returned to the fold as national superintendent of STI in 1922. The Scientific Temperance Federation Papers at the New York Public Library help illuminate the interim period, particularly Stoddard's bitter rivalry with the WCTU and her financial dealings with the Anti-Saloon League.

Hunt and her assistants also left behind a vast amount of published literature, starting with Hunt's own histories of the movement: *A History of the First Decade of the Department of Scientific Temperance Instruction . . . ,* 2nd ed.

---

* Alice M. Guernsey, "Historic Days," *Union Signal,* 6 Jan. 1887.

(Boston: George E. Crosby, 1891), and *An Epoch of the Nineteenth Century: An Outline of the Work for Scientific Temperance Education in the Public Schools of the United States* (Boston: P. H. Foster, 1897).* Predictably biased and occasionally inaccurate, these books nevertheless contain a wealth of material about Hunt's legislative and enforcement strategies. Her pedagogical theories—and her impatience with so-called Progressive education—appear most distinctly in the *School Physiology Journal,* which Hunt edited and distributed to classroom teachers. Local struggles over STI are documented in the WCTU's *Minutes of the Annual Meeting* and in state WCTU reports, housed at the WCTU's Frances E. Willard Memorial Library in Evanston, Illinois. The library also possesses an unpublished index to the WCTU's national newspaper, the *Union Signal.* Since the *Signal* was edited by allies of Willard, it sheds crucial light upon her split from Hunt and upon the larger ideological rifts within the WCTU. These rifts widened after Hunt's death, as shown by Stoddard's *Scientific Temperance Journal* and the WCTU-approved *Temperance Educational Quarterly.*

For primary material on STI and expert authority, I relied heavily upon the Wilbur O. Atwater Papers, Olin Library, Wesleyan University. Atwater corresponded with a wide range of protagonists in the struggle over alcohol education, including physiologists, clinical doctors, school officials, ministers, and WCTU activists. The scientific case against STI was laid out most comprehensively in *Physiological Aspects of the Liquor Problem,* ed. John S. Billings, 2 vols. (Boston: Houghton Mifflin, 1903), while its scientific defenders published most often in the *Quarterly Journal of Inebriety.* For the post-Hunt era, the *Bulletin of the American Academy of Medicine* and the American Public Health Association's *Public Health Papers and Reports* help reveal how hygienists replaced STI with a "social" approach more attuned to liberal pedagogical trends.

The best brief source on educators' attacks upon Scientific Temperance is *Protest against the Bill Unnecessarily Increasing Physiology and Hygiene with Special Reference to the Nature of Alcoholic Drinks and Other Narcotics* (n.p., 1895). As state superintendent of schools in New York, Charles R. Skinner surveyed educational leaders around the country and published their responses in this slim volume. School reports and periodicals also proved invaluable for understanding educators' near-solid opposition to STI. Especially useful were the

---

* Hunt's *History of the First Decade* went through three nearly identical editions between 1889 and 1892. I have drawn almost entirely upon the second edition, published in 1891, which seems to have had a much wider circulation than the other two.

state board of education reports from Massachusetts, New York, and Pennsylvania, along with the *Michigan School Moderator,* the *Ohio Educational Monthly,* and the *Pennsylvania School Journal.* Although typically edited by school officials, these periodicals also helped demonstrate why classroom teachers so often united with administrators in resisting STI. Hunt's own efforts to train "professional" instructors are detailed in the Presidential Papers at American University, site of her failed College of Scientific Temperance.

I also drew upon a large secondary literature about expertise and democracy in the Progressive era, starting with Richard Hofstadter, *Anti-Intellectualism in American Life* (New York: Vintage, 1962). Although Hofstadter set out to examine popular attitudes towards intellectuals, his focus shifted to the intellectuals themselves midway through his book. The same emphasis marks almost all subsequent commentary upon this subject, whatever its interpretive bent—muckraker, pessimist, or celebrant.

Muckrakers emphasize the greed and duplicity of experts, who used new knowledge to subjugate lay citizens and evade questions of social justice. Early texts in this cohort were Mary O. Furner, *Advocacy and Objectivity: A Crisis in the Professionalization of American Social Science, 1865-1905* (Lexington: University of Kentucky Press, 1975), and Burton Bledstein, *The Culture of Professionalism: The Middle Class and the Development of Higher Education in America* (New York: Norton, 1976); a good recent example is JoAnne Brown, *The Definition of a Profession: The Authority of Metaphor in the History of Intelligence Testing, 1890-1930* (Princeton: Princeton University Press, 1992).

Pessimists often share the muckrakers' worries about expertise, but they attribute its expanded authority more to social complexity—that is, to new *conditions* of explanation—than to the stratagems and wiles of the experts themselves. Their theoretical lineage starts with Robert Wiebe, *The Search for Order, 1877-1920* (New York: Hill and Wang, 1967), and stretches through Thomas L. Haskell, *The Emergence of Professional Social Science: The American Social Science Association and the Nineteenth-Century Crisis of Authority* (Urbana: University of Illinois Press, 1977), to Robert Wiebe, *Self-Rule: A Cultural History of American Democracy* (Chicago: University of Chicago Press, 1995).

Celebrants are a more mixed breed. Acknowledging the elitist features of Progressive-era expertise, they refuse to tar the entire project with this negative brush. Instead, they emphasize the humane, democratic impulses of two intersecting expert spheres: female reformers and liberal intellectuals. On women reformers, I have been most influenced by Robyn Muncy, *Building a Female Dominion in American Reform, 1890-1935* (New York: Oxford University Press, 1991), and Kathryn Kish Sklar, *Florence Kelley and the*

Nation's Work: The Rise of Women's Political Culture, 1830-1900 (New Haven: Yale University Press, 1995); on liberal intellectuals, by Robert Westbrook, *John Dewey and American Democracy* (Ithaca, N.Y.: Cornell University Press, 1991), and Thomas Bender, *Intellect and Public Life: Essays on the Social History of Academic Intellectuals in the United States* (Baltimore: Johns Hopkins University, 1993). Whereas Hofstadter viewed intellectual and popular authority as locked in a perennial war, this literature argues that some early twentieth-century experts tried to reconcile them. Yet it echoes Hofstadter's focus upon the intellectuals themselves, thereby blinding us to the ways that laypeople harnessed expert knowledge to their own purposes. An important exception is Brian Balogh, *Chain Reaction: Expert Debate and Public Participation in the Development of American Commercial Nuclear Power* (Cambridge, Eng.: Cambridge University Press, 1991), which examines the post–World War Two period rather than the Progressive era.

With respect to individual professions, I drew most heavily upon the literatures about medicine and education. On the changing role and definition of "science" in medicine, the best single source is still *The Therapeutic Revolution: Essays in the Social History of American Medicine,* ed. Morris J. Vogel and Charles E. Rosenberg (Philadelphia: University of Pennsylvania Press, 1979). For subspecialties within the developing scientific nexus, I drew upon John Duffy, *The Sanitarians: A History of American Public Health* (Urbana: University of Illinois Press, 1990); W. Bruce Fye, *The Development of American Physiology: Scientific Medicine in the Nineteenth Century* (Baltimore: Johns Hopkins University Press, 1987); and Gerald Grob, *Mental Illness and American Society, 1875-1940* (Princeton: Princeton University Press, 1983). Like the general literature about turn-of-the-century expertise, these works are much clearer about why physicians rallied around the scientific ethos than about why laypeople embraced it. In "The Struggle for Ignorance about Alcohol: American Physiologists, Wilbur Olin Atwater, and the Woman's Christian Temperance Union," *Bulletin of the History of Medicine* 64 (1900): 366-92, Philip J. Pauly depicts Atwater's clash with Mary Hunt as a battle over the "public meaning" of physiology, pitting the new laboratory-based science against an older "Christian physiology." I see the struggle as occurring within a shared definition of science, broad enough to accommodate any number of competing explanations and interpretations.

The standard texts on the role of expertise in public education are still David Tyack, *The One Best System: A History of American Urban Education* (Cambridge: Harvard University Press, 1974), and David Tyack and Elisabeth Hansot, *Managers of Virtue: Public School Leadership in America, 1820-1980* (New York: Basic, 1982), which describe how a new generation of scholars and superintendents—the Educational Trust—came to dominate twentieth-century schooling. My understanding of debates among these educators owes much to Herbert M. Kliebard, *The Struggle for the American Curriculum, 1893-1958* (Boston: Routledge and Kegan Paul, 1986). Kliebard identified four different "interest groups" of educators—humanists, developmentalists, social meliorists, and social efficiency advocates—who did battle over the curriculum. Like Tyack, however, he downplayed popular influences upon it: the curriculum was largely forged by *professional* "interest groups," not by large bodies of lay citizens.

Over the past fifteen years, a spate of so-called pluralist accounts—including several coauthored by Tyack himself—have sought to revise this insular image of the American curriculum, showing how citizen groups demanded new courses like health, domestic science, and vocational training. See, e.g., Jeffrey Mirel, *The Rise and Fall of an Urban School System* (Ann Arbor: University of Michigan Press, 1993); Paul E. Peterson, *The Politics of School Reform, 1870-1940* (Chicago: University of Chicago Press, 1985); William J. Reese, *Power and the Promise of School Reform: Grassroots Movements during the Progressive Era* (Boston: Routledge and Kegan Paul, 1986); and David Tyack, et al., *Public Schools in Hard Times: The Great Depression and Recent Years* (Cambridge, Mass.: Harvard University Press, 1984). But even these complex accounts focus mostly upon the general outline of curriculum, not upon its specific content. Business and taxpayer associations sought to constrict the new electives, while labor and women's groups rallied to their defense, yet none of these laypeople retained an active, ongoing voice in determining the day-to-day substance of instruction. Once the course schedule was set, in other words, they went back to the sidelines to celebrate their triumph—or to lick their wounds. A notable exception is David Tyack, et al., *Law and the Shaping of Public Education, 1785-1954* (Madison: University of Wisconsin Press, 1987), which shows how a variety of lay groups—including the WCTU—continuously shaped what children learned.

Perhaps the most important influence upon the curriculum, however, was the classroom teacher. Long ignored by historians, teachers have attracted renewed attention in recent years. A good sample of this scholarship is *American Teachers: Histories of a Profession at Work*, ed. Donald Warren (New York: Macmillan, 1989). I was also influenced by Marjorie Murphy, *Blackboard Unions: The AFT and the NEA, 1900-1980* (Ithaca, N.Y.: Cornell University Press, 1990), although I departed significantly from her interpretation. Murphy argues that administrators used "professionalism" to drive a wedge between teachers and "the community." On a subject like Scientific Temperance, however, communities were often bitterly divided. So teachers retreated into the professional citadel, not to appease their administrative superiors but to avoid an endless torrent of lay demands.

Finally, I also drew upon a burgeoning literature about temperance and prohibition in the United States. Echoing Hofstadter's notorious critique of the prohibition movement as a "rural-evangelical virus," most scholarship before the 1960s dismissed temperance as the status-anxious squeal of a declining, "old" middle class. Since then, historians have almost completely reversed this formula. Temperance took root among the urban "new" middle class of businessmen and professionals, as noted initially by James H. Timberlake, *Prohibition and the Progressive Movement, 1900-1920* (Cambridge: Harvard University Press, 1963), and more recently by K. Austin Kerr, *Organized for Prohibition: A New History of the Anti-Saloon League* (New Haven: Yale University Press, 1985). No longer ridiculed as the provincial domain of blue-nosed prudes, meanwhile, the Woman's Christian Temperance Union has been reinterpreted as a seedbed of modern feminism. The crucial texts in this interpretation are both by Ruth Bordin: *Woman and Temperance: The Quest for Power and Liberty* (Philadelphia: Temple University Press, 1981), and *Frances Willard: A Biography* (Chapel Hill: University of North Carolina Press, 1986). More recently, Ian Tyrrell has stressed the WCTU's attitudes towards race and imperialism in *Woman's World/Woman's Empire: The Woman's Christian Temperance Union in International Perspective* (Chapel Hill: University of North Carolina Press, 1991).

By contrast, my own perspective emphasizes the ideological divisions between liberals and conservatives *within* the WCTU. This rift mirrored many of the era's religious controversies, which are ably discussed in Ferenc Morton Szasz, *The Divided Mind of Protestant America, 1880-1930* (University, Ala.: University of Alabama Press, 1982). Also useful were William R. Hutchinson, *The Modernist Impulse in American Protestantism* (Cambridge: Harvard University Press, 1976), and Susan Curtis, *A Consuming Faith: The Social Gospel and Modern American Culture* (Baltimore: Johns Hopkins University Press, 1991). The classic account of turn-of-the-century American culture is T. J. Jackson Lears, *No Place of Grace: Antimodernism and the Transformation of American Culture, 1880-1920* (New York: Pantheon, 1981), which helped me situate Mary Hunt and her assistants within the broader American debate about free will and individuality. Like William James and other elite "antimodernists," Hunt worried about the erosion of personal autonomy. But she shared none of their pessimism, insisting that each American could freely choose to cast away alcohol. That was the bedrock belief of Scientific Temperance Instruction, an old-fashioned faith in newfangled clothing.

INDEX